T0335810

IS MAN TO SURVIVE SCIENCE?

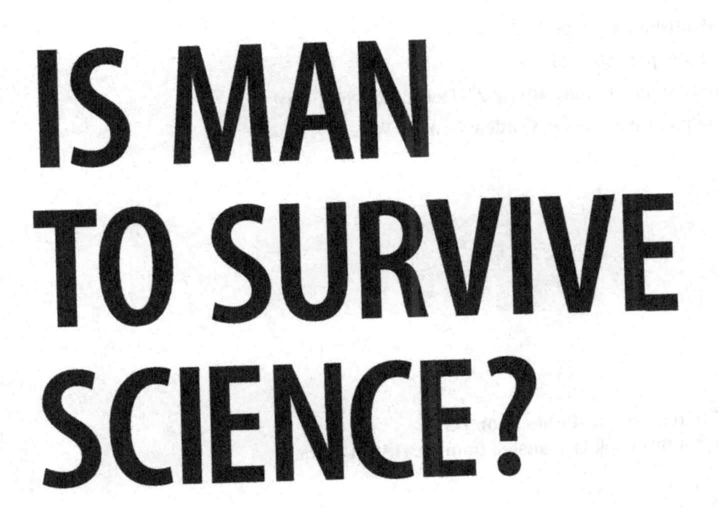

Jean-Pierre Fillard
University of Montpellier II, France

NEW JERSEY • LONDON • SINGAPORE • BEIJING • SHANGHAI • HONG KONG • TAIPEI • CHENNAI

Published by

World Scientific Publishing Co. Pte. Ltd.
5 Toh Tuck Link, Singapore 596224
USA office: 27 Warren Street, Suite 401-402, Hackensack, NJ 07601
UK office: 57 Shelton Street, Covent Garden, London WC2H 9HE

British Library Cataloguing-in-Publication Data
A catalogue record for this book is available from the British Library.

IS MAN TO SURVIVE SCIENCE?

ISBN 978-981-4644-40-2
ISBN 978-981-4644-41-9 (pbk)

In-house Editor: Song Yu

Typeset by Stallion Press
Email: enquiries@stallionpress.com

Printed in Singapore

Foreword

Présenter le remarquable ouvrage du Professeur Jean-Pierre Fillard est pour moi un très grand plaisir. A cela participe bien sûr notre vieille amitié, qui remonte à la fin de nos études secondaires dans un lycée d'Alger, il y a près de 60 ans. Mais si l'amitié donne une tonalité particulière à un jugement, elle n'en exclut pas la lucidité et l'objectivité. Or le livre qui nous est proposé est remarquable à plusieurs égards, par le fond et par la forme.

Avec une maestria étonnante, Jean-Pierre Fillard dresse d'abord, dans «l'Homme survivra-t-il à la Science?», un état des lieux époustouflant sur les connaissances actuelles et sur l'évolution récente des «sciences dures», dont il est spécialiste comme physicien, mais également de la biologie et de la médecine, toutes deux qualifiées, parfois et curieusement, de «sciences molles», peut-être parce que leur sujet, la vie, est éminemment variable.

Les domaines abordés sont nombreux et variés puisqu'ils vont de la physique, à l'électronique, à la robotique, à l'informatique et aux nanotechnologies. Tous ces sujets sont traités avec une maîtrise reflétant une vaste compétence scientifique mais aussi une érudition étonnante dont témoigne une bibliographie faisant références aux plus récentes données. Chaque fois, l'auteur sait faire la part de la science et de la technologie, en soulignant combien elles sont indissociables, chacune d'entre elles permettant tour à tour les progrès les plus significatifs. Ces progrès scientifiques et techniques, le plus souvent très spectaculaires, vont d'ailleurs en s'accélérant puisque, par exemple, moins de 80 ans séparent l' «Eole» puis le «Zéphyr» de Clément Ader, plafonnant à quelques centimètres d'altitude, et le vol inaugural du Concorde à plus de 10.000 mètres d'altitude et à une vitesse de Mach 2. C'est précisément cette accélération fabuleuse des connaissances et des progrès techniques qui en découlent qui amènent à se poser la question de la place et de l'évolution de l'homme dans un monde aussi profondément changeant.

Les mêmes commentaires s'appliquent aux chapitres de l'ouvrage concernant la Biologie et la Médecine, mais dont la qualité est d'autant plus remarquable que l'auteur n'est plus directement ici dans son domaine de compétence. Il nous présente pourtant un remarquable panorama des progrès biologiques et médicaux, dans des voies aussi variées que la génétique, le clonage, les neurosciences, la pensée ou l'intelligence. Dans tous les cas, on voit se dégager la profonde convergence et l'évidente complémentarité des «sciences dures», de la Biologie et de la Médecine. Cette complémentarité est pourtant trop fréquemment méconnue ou même ignorée — souvent en raison de l'esprit de chapelle de certains spécialistes — et je ne résiste pas à évoquer ici le souvenir d'un cercle dont mes parents, avant 1962, étaient membres à Alger, le Cercle Franco-Musulman, dont la devise était une phrase du philosophe Paul Valéry «enrichissons-nous de nos mutuelles différences»...

Face à cette montée brutale, à cette véritable explosion continue des connaissances, comment l'homme pourra-t-il évoluer et quelle sera sa place dans un monde en perpétuel changement? C'est la question majeure qu'aborde Jean-Pierre Fillard, avec d'autant plus de courage que cette question est éminemment difficile, notamment parce qu'elle touche aussi bien à la philosophie et à la morale qu'à la prospective. Certes, les schémas que l'on peut imaginer aujourd'hui ne se réaliseront peut-être jamais et nous feront sourire dans quelques années. Mais quelle qu'en soit la difficulté, il n'est pas moins essentiel de savoir prendre du recul, de réfléchir et d'imaginer ou de prévoir l'avenir. Aidés ou poussés par les progrès de la science, irons-nous vers l'apparition d' «hommes transhumains» ou d'«hommes zéro défaut» produits par des clonages ou des filtres génétiques, ce qui évoque bien sûr la perspective effrayante de l'eugénisme. Sans aller jusqu'à ces extrêmes, l'homme saura-t-il évoluer pour ne pas disparaître et sauver l'essentiel de son essence matérielle mais aussi intellectuelle et morale? Nul ne peut répondre à ces questions mais il est fondamental de savoir les poser et c'est cette réflexion que nous apporte l'ouvrage de Jean-Pierre Fillard

Passionnant par le fond, ce livre est également remarquable par la forme. Les sujets les plus complexes sont présentés avec une parfaite simplicité qui donne au lecteur le sentiment — peut-être injustifié parfois... — d'avoir tout compris, mais cette volonté de « faire simple » n'altère en rien la rigueur de l'exposé. Celui-ci est conduit d'une manière naturelle, presque chaleureuse et, en le lisant, on a l'impression d'entendre Jean-Pierre Fillard dont les qualités de conférencier ne sont plus à vanter.

«L' Homme survivra-t-il à la Science?» est donc un livre passionnant, par les données scientifiques qu'il présente et par les pistes de réflexion qu'il ouvre. Mais c'est aussi, et c'est ce qui en fait également la valeur, un livre facile et agréable à lire.

Meylan 25 Janvier 2013

Professeur Pierre Ambroise-Thomas †
Président honoraire et membre de
l'Académie nationale de Médecine.
Membre de l'Académie nationale de Pharmacie.

This foreword was kept in French in respect of my old mate, Pierre, who passed away recently.

Jean-Pierre Fillard

Table of Contents

Preliminaries

Before everything else, some words seem to be essential in order to get to know each other.

Where do I come from?

First of all, I do have to briefly introduce myself to the reader who has never heard of me. I was (scientifically) born in the late fifties and, at that moment, the old conceptions of our world were disappearing to give birth to modernity.

Physics was the cherished science and generated a bunch of new scientific topics which were strongly divergent, enthusiastic and futuristic. Each of them was developing from itself and in its own direction. In some ways, the times were at a real scientific "Far West", with the triumphant nuclear physics, but also emerging new promises.

Space was open, laser was just born, information was a theory, computer science "lay in a cradle", transmission and television were in the starting blocks, whereas mathematics obviously turned "modern"; however, last but not least, the very revolution will arise from the transistor. What a rupture!

At that time also, a marked scientific priority was given to "hard" sciences. "Soft" ones (?), devoted to nature studies and observation, were in the background; biology, for instance, was considered (I surely overrate it) as a domain for "butterfly hunters"! The initial investigations on the DNA certainly did not occupy the headlines of the newspapers. Things will change and reverse later when "hard" sciences will get tired, giving birth to a strenuous baby, I mean: Instrumentation.

From there on, spectrometers, microscopes, analyzers, scanners, imaging systems, CATs, MRIs, and of course the unavoidable computer, will all become precise, powerful, adapted, and almost intelligent; they will provide the research with an unexpected dimension. Since then the move was such that diverging sciences will get closer and closer to the point that they tend to become now strongly associative.

Let us come back to the fifties; it was then, in this stimulating atmosphere, that I had the great opportunity to meet a prominent teacher, Professor A. Blanc-Lapierre[1] who later became president of the French National Academy of Science. It is a pleasure for me to greet him for allowing me to share his enthusiasm for Physics studies and providing me with a basic scientific culture in nuclear physics and quantum mechanics.

However, my preliminary works were orientated towards the field of solid-state physics which attempted to theoretically

[1] http://www.supelec.fr/actu/Blanc-Lapierre.PDF

strengthen the new experimental discovery of the transistor effect which just came to fruition. This initial work leads me to experimentally reveal the splitting of the "conduction band" in Copper Oxide crystals.

The following was more tragic because we were in 1962 and I had to leave my birth land to migrate towards a hostile homeland. This is mainly because of this hostility that I came back soon in Algiers and engaged in a new research with Ecole Normale Supérieure (Paris) in the field of organic semi-conductors. This was, then, rather unconventional and daredevil, but I got through successfully to the point that I defended a thesis later (*cum laude*) on the semi-conducting properties of Cu-Phtalocyanine thin films, in the new University of Montpellier (USTL[2]) which had just been created. Some 20 years later I was told that a young Japanese researcher was to extend my work to further conclusions!

Then I had to head my own laboratory with some young students eager to struggle with science. The following 30 years were devoted to understand the behavior of the electrons in semiconductor materials when they meet defects such as chemical impurities or structural disorders. Light, heat, electric fields interact to trap, free, recombine them, and emit light, thus perturbing the operations of electronic components.

At that time people were looking for a successor to King Silicium for more performances and the main challenger was Gallium Arsenide. During 10 to 15 years there has been an incredible worldwide activity to master this material.

[2]Université des Sciences et Techniques du Languedoc.

Thousands of researchers were involved in new theories or experimental approaches and any issue of the prestigious *Physical Review* dedicated at least one paper to the subject. Then, step by step, improvement after improvement, the quality of the crystals became sharp enough to render possible to elaborate performing Integrated Circuits in the domain of the High Frequencies, thus giving rise to GPS, mobile phones, TV satellites, flat screens, intercontinental optical fibers ... and also Internet.

A recurring defect was named EL2 and during some 15 years, streams of ink flowed to give a satisfying explanation of the nature of this perturbing flaw. Was it a chemical interstitial atom? Was it a local vacancy or a combination of them? The answer never was clear. In spite of what, this difficulty has been empirically overcome by technology and (almost) nobody cared any longer of EL2. This story shows that if Science is in many cases required for the things to progress, nevertheless technology often finds its way (when shown) by itself.

During this term, we were among the first people to trust in digital imaging. With a Japanese colleague (Professor T. Ogawa since deceased) we imagined a Laser Scanning Tomography (LST) which made it possible, in the bulk material, to visualize particles as tiny as the nanometer scale and we invented a new word: "Nanoscopy".[3] These efforts were

---

[3]Near Field Optics and Nanoscopy, Jean-Pierre Fillard, World Scientific, Singapore, 1996.

comforted by the advent of the AFM[4] invented in Stanford by Calvin Quate who has been with our University as a visiting professor some years before.

This new kind of imaging takes place in the microscope's family which is called "Scanning Probe Microscopy". It works as the "pick-up" of my youth: a tiny and sharp needle is carried at the tip of a small silicon arm and is brought carefully close to the surface of the sample; when the contact is near to be reached some proximity forces arise which repels the needle thus bending the lever. This weak bending is optically detected. Scanning the sample with this equipment makes it possible for the computer to generate a quantitative image of the relief at a nanoscale.

Such instruments (as Scanning Tunnel Microscope) make up an extraordinary mean of investigation in the nanoworld not only for getting images but also to carry atoms, one after another one, from place to place, thus giving rise to individual molecular constructions the physical laws of which leads to a new chemistry. But, anyhow, the nanoworld remains as difficult to get access in as the outer space is.

It also soon appeared that such nanotools could be optical, the tip of the microscope being transformed into a nanorobot capable of directing light to and from a nanotarget. The experiment works well and it makes it feasible to use the silicon arm as a light guide[5]! From that time on, many other tentative explorations in the field have been carried out.

---

[4] Atomic Force Microscopy.
[5] Ultra Microscopy, 61 85 (1995) to 71 231 (1998).

All of this gives, in a summary, what my prospects were during those 30 years. From this long period I remember that the "job" for a researcher is rather unusual, and it is for above average people. Science cannot be won without the researcher having explored, doubted, imagined, and checked. May be, this researcher could realize he was wrong and somebody else may re-study the problem with another mind. Success must be considered with a reservation taking into account that truth could be incomplete or temporary. A friend of mine told me once: "I appreciate when you say: I don't know!"

It stems from all the above that I am in no way a biologist, neither a robotician, nor a computer science specialist; however I still remain a simple, inquiring and objective physicist. But this restriction is not so dramatic at the moment, considering how drastically things have changed during the last decade. Since then information has become accessible, handy, profuse and swift and it turns out that even a non-specialist, as I am, can easily get a valuable idea of what is going on in a different field.

Many things which are today considered as impossible would likely be commonplace tomorrow. Claude Allègre did say incautiously but with due reason in his excellent book[6] in 2009: "One does not know how to synthesize life" but Craig Venter did it since at the cell scale and it is just a very beginning!

The evolution of the Science world can be sketched following three global and entangled trends:

Convergence — Acceleration — Applications.

---

[6]Claude Allègre, "La Science est le défi du XXIe siècle", Plon Ed., 2009.

Each of them provides means for the others to progress.

— Convergence: Biologists now need to deal with software. Chemists require quantum physics to understand individual atomic processes, and Roboticians are called for help by the doctor to artificially copy a knuckle, and so on.
— Acceleration: the speed of transfer of the knowledge gets faster giving tremendous efficiency in the scientific challenge.
— Applications: every result in the research advancement immediately leads to a serviceable conclusion in the form of an apparatus or an instrument which immediately gives rise to new performing investigations, and so on.

To give an illustration of the evolution, my own ancient lab, for a while, no longer cares about Gallium Arsenide crystals. My previous students, now in the business, collaborate in a team with doctors, chemists, software specialists to study biological cell luminescence properties under nano-MRI imaging! Quite a revolution!

Another epoch, another set of manners, another set of nanoscopies!

Then, as a conclusion, I feel perfectly authorized to give my careful opinion about stem cells, genomics or super computers as well! That I will do in this book.

Why such a Book?

Some time ago, a friend of mine gave me a short excerpt of a book he was reading. The title is "La Mort de la

Mort"[7] (The Death of the Death). It was produced by a doctor deeply involved in the business of genomics. This enticed me to get the book and read it too. The developed philosophy was that the progresses in the general scientific knowledge have become so fast that fantastic conclusions are to be soon reached which will overturn our ways of living with very short delays. More especially genomics, stem cell studies, computers could make it possible to largely improve the medicine to that point that living 150 years or much longer could become soon a common reality, thus leading to questions humanity will have to face without any possible escape. It seemed to me that the arguments were absolutely serious and were worth being considered to the point that writing a book would be an exciting challenge.

One can find today many publications dealing, under different angles, with the subject we are busy with. It can be observed in many cases that authors often use thoughtlessly mathematical expressions or the word "exponential" or the prefix "nano" which sound convenient in a scientific context, but are often in any way not justified.[8] Such enthusiastic outbreaks are prejudicial to the credibility of the purpose, however, we have not to dwell on the matter but rather look forward.

---

[7] Laurent Alexandre, La mort de la mort, JC Lattès Ed., 2011.

[8] Even Ray Kurzweil had this sentence which could horrify a beginner in mathematics: "exponential trends do reach an asymptote!"

It seems profitable to me to draw a lesson that, not with the aim of striking a balance or forecasting something, this would be presumptuous and vain because nothing stays final, but rather we should attempt to open windows of thinking over a realistically possible future (if not likely). This is an intimate or collective future which is to happen soon and in some ways engraved in our destiny. We will certainly accept these transgressions, our children will know about them and our grandchildren will surely have to live in. Humanity will certainly not get undamaged out of this new "shock of the future". We might as well get seriously prepared to approach it.

In short, the book is organized in the following chapters (as a matter of fact you may note the many? in the writing):

– Part one

What are the undisputed facts that might be involved?

* Chapter 1

How did we get to that point? History is a rather long process from the primitive ages where various steps of the technical evolution have piled up, giving rise to the present status of "Progress".

* Chapter 2

A new start has been obtained when microelectronic physical devices were developed leading to new fields of "nanotechnologies" that now extend to the biological domain.

* Chapter 3

Information and software science have benefited a huge extension towards robotics, artificial intelligence and simulated virtual worlds.

* Chapter 4

The biological sciences were for a long time limited to only an instrument: the optical microscope. But now new facilities merged that gave a kick to deeper and deeper investigations in the cell domain and genomics.

* Chapter 5

Brain was, from the beginning, an unfathomable mystery. The first valuable lights were only thrown during the last decade where cognitive sciences began to appear, leading to preliminary but fantastic and provisional conclusions.

Part two:
What moral is to be expected, hope or despair?

* Chapter 6

What kind of questions could be reasonably asked? Is Google to replace Big Brother? Have we already engaged the changes to lengthen our beings?

* Chapter 7

How should we consider the scientific research in this new framework? What about its impact on our daily life and what place would be given to the computer?

* Chapter 8

The Human is undoubtedly to be the target of technical assaults arising from several directions: gene manipulations, electronic implants, even cerebral transfers into a computer! Would the machine rationality become prevailing?

* Chapter 9

The anticipated possible modifications of our very substance could be provided through the advent of "zero defects" artificial babies. From that on, one could ask the question: Is the Human being still necessary … or even desirable?

* Chapter 10

Then the ultimate question arises: what about God in such an upheaval? Would it be necessary to set up a computer assisted "neurotheology" to replace the essential idea of God?

Introduction

Futurology finds its source in the past and present history. What is bound to happen would be a direct and logical (or not) extension of what does exist; one calls it determinism. Of course, this should be completed with a touch of imponderable without which nothing could take place. Our recent history looks fascinating because the move gets dreadfully accelerated. A century ago, a man's life was not long enough for the "progress" to be perceived whereas, nowadays, important changes are to be noticed on a yearly basis. Our children will certainly see the very fast phases of the evolution, they will see the steps, the direction too, and they will have to adapt.

Millions of years were required for the life on earth to select a species smarter than others. "*Homo habilis*" was a naked man hardly standing on his legs. Then things got much faster for "*Homo erectus*" shortly followed by "*Homo sapiens*". All along the Neolithic times the steps of "elementary cognition" were shorter and shorter.

Recent publications referring to the subject suggested by the title display a rather large diversity of accesses: biologists, software specialists, geneticists, doctors, artificial intelligence researchers, even philosophers, each of them offering his own short-term extrapolations out of his respective science. All of them converge towards a similar conviction: the expected changes should be powerful and should occur fast.

The aims of this book are to take stock of the questions, to give the reader a global view on all these promises which certainly are to overthrow our ways of living and even our fundamental concepts, and this, no doubt, in a close future. We all are concerned for the better but also may be the worst. We all will have to, individually or collectively, bear this unavoidable evolution.

At all times, the evolution scheme always was the same: at the outset a distinguished scientist proposes an explanation to any phenomenon. For instance, Archimedes, in his bath, suddenly screams "Yeah! I understand, I have discovered my principle! Eureka!" He shouts out the news to everybody around. Then comes a technician who says "well, well, ... that may help to do something! I will make a machine, any device or instrument." In that manner he, for instance, will build a submarine. Then the human society takes this opportunity and decides to use this submarine immediately to go and see the fishes ... or, better, to make war! It is really the human society with its unquenchable appetites, which decides to use the innovation and the way to use it.

The only difference between yesterday and today is that, yesterday Archimedes was alone (I guess) in his bath, whereas today thousands (millions) of technicians immediately are busy with proposing new gizmos to our society which collapse under solicitations. Delays are drastically reduced and we are overwhelmed with innovations, all very attractive (and also sometimes disturbing), all to be forgotten when the following one arises.

I would in no case be at the origin of nightmares (some pessimists reproached me with that) but simply show the present realities and ask some questions, not to give answers unfortunately! The counterpart of that is a huge expectation for unthinkable benefits and that is what we are to remember.

The subject is boundless, complex, and mobile. I just tried to be simple by assembling, may be in bulk, information that came from the current daily events. However, as I was writing this text many details could have deeply changed although the references I use are almost exclusively from the recent years. In this order I also used many links to Internet which could be volatile.

The present intents are in no ways to predict a catastrophic future such as a nuclear accident, a destructive virus, a climate reheating or a world economic crisis. Nothing at all, it is more serious because we bear collectively with us the likely catastrophe; it is part of our destiny. If, by chance of the History, this would not have to happen then a much larger event should have arrived which should have blocked it. We have to accept our destiny.

The theory of evolution, our biological programming

The story of the evolution of the species is a very old one but Lamarck paved the way with the "transformism", then Darwin came with his "evolutionism" theory which appeared to impose on us (somewhat hardly). These postulate that species do not stand still; instead they evolve, diversify or even naturally disappear.

These changes among populations have come slowly from generation to generation as a result of natural selection or of constraints from the surrounding environment (ecosystem). This theory got sharper with the genetics discovered by Gregor Mendel[9] and is now universally accepted.

This is available for all species; Jacques Monod[10] would say in a metaphoric way: "what is available for bacteria is as well true for elephants". This particular property of adaptability has been verified: one says that in the year 1971 a lizard species (so-called "ruin lizard") has been introduced in an isolated Adriatic island and left here alone. Thirty-six years later, these lizards were again studied by a biologist who found that deep changes had taken place in their morphology and their insectivore habits, they became herbivore! Evolution is then a phenomenon that can be noted not only on cellular organisms but also on developed species.[11]

---

[9]Grégor Mendel, Versuche über Pflanzenhybriden, Verhandlungen des naturforschenden Vereines in Brunn, 1866.

[10]Jacques Monod, Le hasard et la nécessité, Le Seuil, 1973.

[11]One can also mention the Galapagos Islands iguana that became a sea animal feeding itself with algae.

Evolution does not take place, contrary to the usual opinion, towards a final goal of perfection (Human being?), but only towards a better adaptation to the surrounding environment and its possible changes. The species simply disappear if they do not adapt fast enough or well enough. The individual has not been given any importance from a strictly biological point of view, what only matters is the survival of the species.

Then death is necessary for the evolution to proceed. The ability to reach the age of puberty is the only limitation to the species perpetuation. The body that does not change to adapt, gradually drops[12]; this is the role of the gene intermixing.

Constraints to the natural selection influence genetics and the induced changes migrate through the heredity. From that idea some deduced that "man is coming from the monkey". This claim was accepted for a long time, but today it is put into question: man and monkey might be two close but separated branches of the phylogenetic tree of the life.

During the XIX century, the idea that man could evolve (favorably?) under the influence of the civilizations gave rise to the theories of the Eugenism, the achieved ideal model being (of course) the occidental civilization.

It can be taken for sure that the first priority of the evolution laws is focused on a single target: to ensure the survival of the species even when introducing the required adaptations induced by a hostile environment. This implies an

---

[12]This is the so-called "Red Queen" theory (*cf.* Lewis Caroll, Alice in the wonderland).

intense reproductive activity. The cards have to be "reshuffled" vigorously!

Primitive men were animal-like and relatively weakly protected from the aggressive ambient. They were animals among other animals under a strict application of the Darwin theory. Child and maternal mortality rate was frightening despite the fact that the biological organism incorporated a strong defense-in-depth at the cellular level within this early age range.

Nature is so made that it was vital to, as far as possible, protect the children until the puberty time is reached, and then they have the duty to procreate as fast and as generously as possible. Those who were not strong enough to survive disappeared and did not transmit their genes to the following generations. This mechanism provided a kind of "genetic filter" which values the favorable mutations against the harmful ones.

Around the age of 18, the survivor must have accomplished his duties concerning Nature and so nature began to give up. Then the aging came thus lowering his capacity to survive; following a more or less extended grace period depending on the circumstances and the individual. In any case, the mean life hardly extended to 30 years.

When "Homo Sapiens" began to find housing, to make fire, grow animals and cultivate the ground, the living conditions greatly improved which makes it the first deviation to the natural basic biological program. Life expectancy lengthened (essentially the grace period) but puberty still remained at the same age[13] whereas the genome began to worsen. The filter got less strict.

From these times, Man never stopped thwarting the established natural laws by adapting, this time, the ambient to the species and no longer the opposite. Hopefully, Man enjoys a reproduction period especially large!

The operational principle of the primitive man still remains valuable for the modern one, it is engraved in our genes: from age 18 on, the auditory performances decrease in the high frequency range, sight also diminishes later, then some get bald ... and lose their sexual health as well!

Nevertheless, the changes in the framework of our lives and also the living considerably limited the chances of child deaths thus protecting fragile children. In the year 1800, 50% of the children died before reaching an age of 7; today a premature or frail baby does have the right (and even the obligation) to live and he will be able to also procreate. This tendency is, however, compensated by the common practice of assisted abortion.

The genetic filter no longer applies thus leading to pollution in our global "genetic heritage". This can be observed especially in the brain's capacity of the adults. We do have infringed, not intentionally, upon the fundamental laws of the Nature but the biological clock still remains unchanged. It is by chance that we get old! The immortality of some, rare,

[13]It has recently been observed that the mean age for girl's puberty may have turned younger now, due to the modern way of life.

species remains accompanied by a blockage of their own evolution.[14]

Geneticists have serious concerns about this problem of heritage, the only solution of which is an active control of the gene therapy in order to put things right before it is too late. Hopefully, this shift is very slow!

Then how to carry out a stretching of the mean life, a stretching of the grace period beyond the normal cycles, without any decline? We do not know yet. May be delaying (or suppressing) puberty, thus misleading the biological clock and postponing the trigger point of the aging mechanism?

These grace delays are benefitting the living species , such delays greatly depend on the standard of living of the involved species and are usually short for mice, whereas man is more largely favored, but however some turtles can live over 150 years, urchin whales can live for over 200 years, and the olive tree, the all-round world champion, for up to 5000 years. But there are also unlucky species which die soon after reproducing (males especially of course, females can get out of that), for instance, the praying mantis or the salmon. We do not know the exact reason for such differences between species.

Today, concerning the Man, the PID or Pre Implantation Diagnosis (which is still contested in some countries like France) allows selecting the "good" embryo to be implanted for an IVF (*In Vitro* Fertilization). That makes it feasible to

---

[14] Hopefully, because in such a case they should have invaded the whole planet!

create a little brother with the only aim to provide the leukemic older with a qualified bone marrow donor. Not easy to definitely discourage such a practice. Similar extrapolations abound. The genetic lottery only results with winners!

From yesterday to nowadays

Man applauds the scientific progress; would it be for his own sake or for an inescapable step in the Darwinian evolution of the species? The only certainty we have is that cataclysmic changes (for the better but likely also the worst) will inevitably take place in a near future. Whether we want to or not, they are on the way and we are going to accept them.

Man's history is a continuous series of pains, wars and indefensible killings, also a struggle for survival. But as a counterpoint to such miseries a fabulous *thesaurus* of observations, deductions, hypotheses have accumulated across the millennia, transmitted from a generation to the following and we called it Science.

These intellectual achievements do have practical implications (one says "technology") which gave rise to undeniable improvements in our daily life, in our comfort, also in our health particularly. We are talking about Progress even there are no counterexamples existing.

This so-called Progress has not always been accepted easily, it is subject to a natural resilience and some even fear to adopt innovation as long as it is not imposed when coming from outside; a kind of "precautionary principle" in some way. Nowadays such an attitude, when pushed too far, can generate serious economic consequences.

Men do not behave in a homogeneous manner towards Science; scientists are bound to their knowledge although they strive to share it largely. That task is not an easy one. They do stick to their certainties and they are obviously the only ones to project for the future, each one in their own domains. Simple guys will undergo and often have to trust in. This is in no way easy because they often meet self-declared "experts" who are only interested in the financial returns.

However, Progress is likely accepted when its usefulness is obvious through the comfort, the fun or the security it brings with it. The safety of some applications still remains to be testified, so are the Smartphones, universally accepted even if they are certainly not as innocent as we do know they make our children deaf before long.

Reservations often remain strong before accepting a Science which is not understood and which generates a native distrust. One can remember the terror screams when the first heart transplant was announced in 1967 or the instinctive rejection of the Genetically Modified Organisms (GMO) or the stem cell studies. This without mentioning the nuclear plants which everybody condemns without reducing his own electrical consumption!

Time is required in order for the Progress to mature, improve, spread, and contribute in our lives without looking back. Time is an essential parameter to get accustomed. But now time is pressing, the Science is exploding and things get an acceleration never achieved before and hardly manageable.

Let us recall some striking examples of such acceleration:

— Two small centuries were required to move from the Cugnot's "fardier" (first terrestrial motorized vehicle, powered by a surrealistic steam engine!) ... to the moon rover, first lunar vehicle packed with an already obsolete technology.
— Only 80 years (that is to say a man life) elapsed from the first "flight" (some centimeters) of the plane Eole with Clément Ader in 1890 to the first flight of the supersonic Concord in 1969! 80 years, no more, with two world wars in between, that is so short for so fantastic a mutation!

A boy who could have attended the Eole take off could as well, in his old days, have enjoyed a journey at Mach 2 and an altitude of 50,000 feet, sipping his cup of champagne in a real comfort! Is that not fantastic?

Now our thirst of knowledge largely spreads, associated to the idea that it is a source of welfare and comfort of living. This justification through the "lollipop" of the Progress results in a larger involvement of the individual but, on a second thought, is this search for the knowledge the only important and sustainable thing in a man's life? May be it is the remaining trace, the stone left to the monument; even if it is very small and even if it looks needless.

The proportion of people who are involved in this research activity do not stop growing and this is an obvious pledge of evolution even if a long way still remains to go before the mass of men gets conscious of the importance of the thing. Some 50 years ago researchers were not so numerous, such an activity was rather marginal, not well understood; today armies of scientists and technicians of any kind are scraping the various fields of science and technology. They are declared of "public utility"; constantly renewed miracles are expected from them while fearing disastrous novelties they could draw from their Pandora box. Opinions differ.

The researcher's "tsunami" just began, they can be found at any level and in every structure: academics, industrial, medical, economic, universities, of course. Already a galaxy of researchers has sprung up. They are everywhere and are expected to do everything. They are the hope for economic prosperity and achieved welfare in every country. We are living in the expectancy of the future: we are not yet accustomed with iPhone5 that we are eagerly waiting for iPhone6 and crowds are rushing to the shops the day of the first selling in spite of a deterrent pricing!

Our concerns for the future has taken a huge importance in our daily life and this constitutes a religious infringement because in many cases foreseeing is considered as an intention to go against the divine will which is intended to govern our destinies. Christian religions do not firmly oppose to that but Islam does. It considers that such an attitude is truly blasphemous and only Allah will lead our lives. This is to the point that the tense corresponding to the future does not exist in the Arabic language. One says it is the "unaccomplished" which only Allah can afford; then it is vain to try changing anything.

One could dream, within some centuries, of a humanity only concerned with Understanding! The every futile need could be satisfied by the means of a largely productive and automated technology. Human brains could be connected in a net provided with a virtual mass memory! But should such an ideal wisdom be reachable for humans; one says that the man's worst enemy is himself!

What would happen if our scientific rocket still continues to accelerate so brutally? Would we have time enough to understand what is going on?

The new previsionism

Our future is, as a matter of fact, a mystery no one could solve; however, the recently acquired knowledge allows some more or less realistic but basically inevitable prospects. It is no longer a matter of an unbridled imagination neither romantic science fiction but very realistic prospects.

Some visionaries, as Steve Jobs, used to say: "The best way to predict the future is to invent it", that is exactly what we are trying to do here.

Forecasters or "Previsionists" always existed, looking for signs of the evolution of the human destiny. Some predicted a bright future enlightened by the science's advancements. Therefore, Darwin anticipated a spontaneous adaptation of the organisms to the difficulties arising from our environment; others, more pessimistic announced frightening catastrophes ready to come, so Malthus forecast an ineluctable world starvation because of the difficulty to grow food enough for an increasing population. He was obviously right but could not foresee the arrival of the agricultural machinery which completely changed the story. The previsionism gets discredited during long years and we had to wait a lot before the "Future chock" by Alvin Toefler[15] which was a very best-seller in the seventies but is now quite forgotten.

However, this book deals more preferentially with the society framework than with the scientific component. 50 years later one can find in this book some announcements which have been deadlocks but, more importantly, we do not find the new ways which have since developed and appeared relevant.

Nevertheless, predictions sometimes are largely mistaken. Let us recall that, after World War II and the following "Cold War" forecasters used to develop theories of self-destruction of Humanity by the means of an extended use of nuclear weapons. They argued that these new bombs are bound to destroy and let behind definitely uninhabitable countries. These forecasters bet their bottom dollar that life will be no longer bearable in Hiroshima and Nagasaki for the centuries to come and this opinion was, at the moment, universally accepted. They obviously were wrong; 10 years after the

---

[15] Future shock, Alvin Toefler, Bantam Books, 1970.

bombings, these towns were rebuilt and now they have turned into bright cities housing millions of people.

Today, the gurus of the future are back with impressive arguments, especially as everybody is able to observe, over time, floods of new and disturbing elements that Science will not stop bringing to our *thesaurus* of knowledge and this in every kind of domains.

No surprise that these new *pythonesses* are established scientists often coming from two distinct but regularly convergent domains: software science and biology. The credibility of their ravings is such that it became necessary to create a dedicated University and called it "Singularity University" in San Francisco. This business is sponsored by *a priori* serious people: Google, Nokia, NASA and so on. An office of this university is on the verge of being settled in Paris in order to prospect disruptive leaders.

This is by no means a whim from fanatics, it is not a sect. This structure aims at teaching realistic approaches to work and offers new perspectives to corporate executives of the industry or economics, regardless of the discipline, in order to anticipate opportunities to come in the world market, because "business is business" and nothing waits!

Then it is no surprise that Americans took the lead in this futuristic adventure, since they always were fond of large spaces and never hesitated to place the necessary means in their quest for the absolute. One may remember the SDIO (Strategic Defense Initiative Organization) created by Ronald Reagan in 1983. At that time the problem was to outbid USSR aggression in the space conquest and defense. The objective was easily reached. Huge budgets were invested and this also helped obtaining significant advances in various domains such as high power lasers, electronics, communications, active optics and so on ... not to mention an avalanche of technological unexpected spinoffs ... such as the adhesive tape. The American public largely followed, creating a lot of associations very similar to sects which pushed for "star wars".

In the present case, it must be stressed that reasonable forecasting is needed for the companies in order to anticipate the evolutions of the markets and the related techniques. A company like Kodak for instance benefited 20 years ago from a worldwide quasi-monopole in the domain of photographic films and perfectly controlled the evolution of this technical domain. Their forecasters however did not notice the arrival of digital retinas and within a few years the whole company was downed in spite of its efforts to recover and to adapt to the new market. As well, IBM never imagined a possible future for small personal computers; it was only after the dazzling success of the Apple's Macintosh with its mouse, its pointer, its white screen, its icons, and ... and its integrated OS that the giant woke up and, at the end of the story, gave up.

International meetings are held on a regular basis to update on future trends. The last one[16] was ambitiously captioned "Project Immortality 2045". This meeting gathered

[16] Global Future 2045 Congress, Moscow, February 2012. http://.2045.com ou bien http://gf2045.com.

scientists of all kinds but also, more importantly famous billionaires (of the Forbes association) among which the Russian Itskov Dmitry, for providing private opportunities for funding. This community presently has some 10,000 members, which is not to be sneezed at.

The big visionary guru now is Ray Kurzweil,[17] software specialist, economist, philosopher and above all a successful businessman. He is famous for his many books[18] and announces a "singularity" in a near future. This mathematical term refers to a definitive "rupture" in the progress of things and in the case in point a sudden rupture in our ways of living. We shall come back to the meaning of the word a little later; the followers of the predication are called "Singularians". All is not to be ruled out in the audacious proposals they develop but caution must be exercised.

In the old Molière's comedies doctor Diafoirus used to speak Latin in order not to be understood by common people, today it is quite different, the watchword is: communication! Scientists seek to avoid the esoteric language; they should participate towards themselves and the others in such a way that ordinary people can follow. In spite of this they hold their truth alone and in order to have it transmitted they do have very often to simplify, schematize, use images with the standing risk of being simplistic and poorly understood. Forecasters

[17]Some, however, do not hesitate to call him a "high level charlatan". Nevertheless, he was recently hired head of the prospective department at Google which assuredly is not for nothing.

[18]The Singularity is near, Ray Kurzweil, Penguin Books, 2005.

do fall obviously in the same trap and their credibility will strongly depend on their own talent of clear language.

What can already be observed in this recent science impetus is, on the one hand the convergence of the various disciplines towards common conceptions, communications and languages and, on the other hand, the need for serious infringements with respect to usual convictions, long standing customs, even universal moral values which were until then admitted as fundamental, well beyond the only religious laws. All of that is to be reinvented on new basis.

Some are talking about a new Humanity to grow involving "improved", "repaired", even "augmented men" by means of biology or electronically; some others outrightly imagine "post-humans"[19] or "trans-humans" that could artificially be elaborated or "optimized". All of that being reinforced by an impressively long life span and it can become even reachable to suggest a possible immortality. The technical feasibility of such a mutation would be achieved, in their opinion, around the middle of this century. This could appear very suddenly and inevitably.

The program of the "Avatar Project" funded by Itskov Dmitry plans to create, before the year 2035, an "artificial brain" which could be fed by a "human spirit". In the same way, a "humanoid hologram" (*sic.*) could also be proposed which could be absolutely identical with its human counterpart.

---

[19]Jean-Michel Besnier, Demain les posthumains, le futur a-t-il encore besoin de nous? Hachette Littératures, 2010.

Are we to believe in such phantasms? Even though all of these fanciful ideas are not to be rejected outright, the "trans-humanists" often call for the good questions. The frontier between science fiction and the possible science has really become very fuzzy.

Part One

Today, the Facts

Chapter 1

How Did We Get There?

It's a long process, a steady maturation which leads the Nature and us, to our present status. Let us see how things took place and how they brought us where we are since the evolution of our mortal destiny.

Life, death and immortality

The main, question which mobilizes the forecasters, remains: Is our mortal destiny inevitable? Could we not bring to an end this abhorrent practice of trespassing or, at least, further strongly delay the timetable of natural passing and its train of physical decline? Richard Feynman, former Physics Nobel Prize winner, said in 1967: "it is one of the most remarkable things that, in all of the biological sciences, there is no clue as to the necessity of death".

Each has his own magic solution: fiddling with the genome using cellular biochemistry, stem cells, or even electronic implants if not a complete transfer of the "self" into a time proof computer by copy/paste or downloading!

Today, a French doctor, Laurent Alexandre,[1] dares to ask: "Is the first man to live 1000 years already born?" This, of course, is provocative but not so quirky as long as the arguments are undoubtedly serious. He is moreover followed by Professor Miroslav Radman[2] who also says "There already does exist small "critters" which are quasi immortal". Whereas Jean-Marc Lemaitre[3] (in a laboratory close to my town) is able to "resuscitate" 100-year old human cells and he says: "the aging wear has been removed ... one can imagine it could be possible to delete the illnesses related to the old age".

Another possibility is the creation of a human clone and we are on the verge of doing so, but such a process would not apply because, even if the clone were to be very genetically similar, he will never acquire the same education, the same life, the same "experience" and, top of that, the clone, on his own, would certainly look for a personal behavior and get his own "self".

Here are the challenges Science proposes to us and which we are to face soon.

In the olden days, death was familiar to the human society and, if you allow me this bad joke, death was part of the life!

---

[1]L Alexandre, Conference, April 2 2011 — "First death fair", http://philosophies.tv/evenements.php?id=635.

[2]MRadman, Interview: http://www.youtube.com/watch?v=Z23XEyXlyf8.

[3]Jean-Marc Lemaître, Midi Libre 1er Novembre 2011.

We attached very great importance to death and the ceremonies were sometimes magnificent with imposing memorials. The modest funeral hearse crossed the city at the slow pace of the horse and people used to doff their hats and cross themselves along the ride. Religion was to support the transition. People did what they could to play down the event because they knew it was inevitable and Heaven (or Hell) was to welcome the deceased.

People were not attached so much to babies in the early ages, as we usually do today, because they knew they will suffer to see them die before an age of five, which was rather frequent. They said "God calls them back".

Soldiers bravely went to death in big battles where one used sabers, guns and cannons without soul-searching, letting thousands dead in the field and also injuring who were in no better shape. Such suffering did not shock anybody as that belonged to the usual habits of traditional fearless heroism. Nowadays, when a soldier is killed (that is of course obviously deeply sad), the event reaches a nationwide scope, media rushes, one asks why, the heck, war is not fought with robots (and this is on the verge of being done)!

We do have become psychologically fragile, we do no longer accept the death and resignation vanishes. When death strikes unexpectedly, following an accident, a crime or any attack, for instance, or when it reaches a person famous in the media, then an anonymous mob gathers as for a union meeting; balloons, flowers and candles are set up to make the event publically acceptable.

All of this furiously recalls old exorcism ceremonies practiced in order to drive out the evils. A psychological unit is immediately dispatched in order to calm tempers.

Death penalty no more applies even for the worst irredeemable criminals; "rehabilitation" is preferred. One no longer dare pronounce the word, one have better to say "end of life". Would it be the unconscious requirement for an extension, a delay?

Moreover, the formerly common expression "to die of old age" no longer holds. We have to look for another cause of death because the conviction that even an old age cannot be fatal intuitively gets into our brains. One can no longer die of aging; "natural death" has been eradicated. However, we necessarily have to die from something that fits well with the medical philosophy which, by now, rises: if the cause of the disease is conveniently identified and if we have a remedy, then there is no possible death to occur (or, at the very least it can be largely postponed).

One would like to go on a death strike; one refuses death and, to our subconscious self we ask ourselves how is it possible that scientists and technologists did not yet put an end to this terrible achievement. Minds have changed to such an extent that we would be ready to ask for, or even require from science, fast solutions to this immemorial problem. Let us be patient: Science reportedly promises us good news, soon.

From such observations, it appears that many complications and implications will immediately arise if a success (even partial) is to happen, thus largely hampering our ways of existence. Nevertheless, an important stretching of our lives is undoubtedly to occur sooner or later as already did the impressive changes in the mean life expectancy.

However, we have to be careful with the statistics terminology: the mean life expectancy refers to the expected mean lifetime promised to a baby at birth. Then it becomes evident that recent improvements in the basic requirements such as hygiene, food habits, medicine and vaccination, have resulted in a powerful and immediate impact on the child mortality which has rapidly decreased, but does not directly affect the maximum life. It only means that much more people will be able to reach these ages, not necessarily to go further.

According to the reports, in France, the life expectancy rate increased from 48 to 79 years (especially for men) within a century. Today we gain 3 months each year and that is a lot. Obviously, these improvements also have an indirect, least important, consequence onto the maximum life because medical progress is cumulative. When advancing in the ages this progress has a more reduced influence but it still works until the limit.

Immortality, or at least extended life, is an ongoing subject, no longer a phantasmagoria or a paranoid fantasy but rather a practical issue which is to be questioned in a not too distant future.[4]

---

[4]Woody Allen would be pleased who said: "I don't want to achieve immortality through my work, I want to achieve it without dying!"

Many scientists address this subject of death. For example, biologist Miroslav Radman who very seriously tells us: "I have seen (from bacteria) a possibility, a projection, of a possible application to the Man which could drastically change our human civilization." The questions this researcher asks call for attention because they result in deep moral, social, economic, philosophical and of course religious transgressions. Transgressions is the keyword and we are about to observe the first harmful or positive effects in our surroundings today.

For instance, the simple fact of getting cured to prevent or delay death, which looks today obvious, has been considered a moral transgression which has not been accepted easily because it is a direct reaction against the wills of Nature, against natural selection which is to govern all living species.

In the Socrates's philosophy, doctors and more generally, physicists reverse the divine order of the knowledge and their researches are considered unholy. Koran itself refuses such an approach opposed to the divine wills. But Muslims are not alone, one may recall the unrelenting struggle of pastor Timothy Dwight, then President of the very respectable Yale University, against smallpox vaccines; smallpox was at that time a major plague which decimated populations. Nowadays, many sects are generically opposed to medical care, transfusion and so on.

In order to comply with the Progress advancement, should we have, then, to transgress all the implicit laws which accompanied us previously? This is highly possible and I do not know if I have to add "alas", because the positive aspects are not to be neglected.

Accelerated history

Our present civilization is the culmination of an evolution made of multiple little elements which accumulated erratically during centuries, but also decisive stages which strongly affected our way of life. They stake our history as white stones on the road and they have changed Man definitely in his way of living, his behavior, his beliefs, thus affecting the whole progeny. These changes get carved in our way of life without any thought of turning back.

Let us recall some milestones we could actually not call in question because of the advantages they brought us. First of all, historically, it was unquestionably the mastery of fire, some 450,000 years ago which started the scientific engine!

One can hardly imagine the panic the first man to light a fire by hurting two stones has likely thrown out. All the neighborhood surely would have prostrated before him as this discovery was unthinkable. He was, ahead of time, the first "Dr. Folamour"! He had in his hands the means of putting fire everywhere he wanted! The fire we, as animals, were so frightened of! But on the other hand … it was so convenient! We often react the same way nowadays in front of an unexpected but worrying technical novelty. Nevertheless the progress was, then, obvious and the invention has been quickly adopted.

From there on, the creation of settlements and agriculture appeared after 400,000 years of expectation, then writing after 8000 years, followed by the Greek's scientific philosophy (2500 years), then printing, steam engine, electricity and so on. One remarks, in any case, that the delays get shorter and shorter between following events.

Last but not least, the further stage before the avalanche is undoubtedly the transistor, some only 60 years ago when electrons were domesticated in the very bulk of matter. The technology will then be able to make the electron perform stunt prodigies which will turn our lives upside down.

Computer has significantly affected our daily life with the unavoidable Internet and so many consequences, I give up to list, as they have gathered during the 10 last years. Everyone has a direct impact on our life and behavior.

It may also have been observed that, formerly, the changes due to the humans mostly originated in a single man (the Scientist) even if there was a preliminary destination. So, Gutenberg was alone to imagine his ink and movable types. Today, any technical revolution comes from teams of researchers and technicians originating in various fields and which work in a stacked way before leading to a solution. It is a continuous and organized production and no longer an individual and sporadic one.

On a larger scale, Ray Kurzweil assembled from literature a compilation of data put together in a single and amazing diagram (of course logarithmic in order to bring them all in). The *x*-axis is devoted to the time elapsed between two

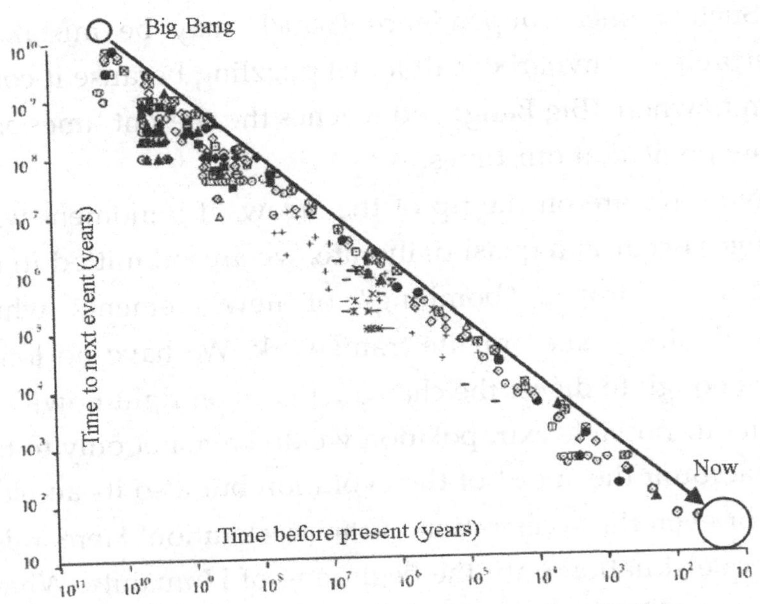

Compilation of significant historical events from Big Bang until today (After Kurzweil[5]). The graph shows the delay between two successive events as a function of the time-lag (inverted scale).

successive events whatever their nature (astronomic, geologic, biologic or human as well) whereas the *y*-axis refers to their date of emergence. It can be clearly concluded that the shifts that, in the beginning, were in the range of million years, now converge towards much shorter values.

This is rather an unexpected result. It is noticeable that the taxonomic point cloud fully complies with a linear law. Then, surprisingly, human events obey a simple natural law arising from the deep ages that should have a hidden meaning because the events we are taking into account belong to fully uncorrelated and drastically different origins.

[5]Ray Kurzweil, The singularity is near p. 19, Penguin Books, 2005.

Such a linear dependence (which may be mistakenly interpreted as hyperbolical) looks puzzling because it comes from nowhere (Big Bang) and reaches the present times as an arrow pointed at our times.

Now we are on the tip of the arrow, at a moment when changes occur in a quasi-daily rate, we are submitted in real time to a proper "bombing" of new elements which, individually, shake our life framework. We have no longer time enough to digest the change, it goes on right now!

In our possible extrapolation we do have not only to take into account the speed of the evolution but also its acceleration or even the acceleration of the acceleration! Here, this is a unique situation until the beginning of Humanity. What is to be found beyond the tip of the arrow, beyond the abscissa wall?

At the other hand of the diagram, the mysterious Big Bang hides the origins of worlds. What did exist before? Was there only a "before"? What is the meaning of all that stuff? The whole history of the Earth inhabitants holds between these two limits.

Nevertheless, today, at the end of the road, our habitat is comfortable, we are usually well heated in winter, we are nicely fed, we light our homes at night, we are able to travel across the whole planet, for no reason often. All the knowledge can be reached with a simple mouse "click", our life expectancy rises, as said, by three months per year.

So "What else?" would say George Clooney sipping his cup of coffee!

Our imprescriptible curiosity, however, bores into ourselves, we need to know what is hidden behind our very existence itself; this mechanism launched for more than 450,000 years has placed us on an infernal slide we do not master, it is our destiny which builds up, before us, uninterrupted.

Tool, machine, system, ... and what else?

From the Neolithic ages to present times, the steps of our "elementary cognition" are becoming shorter and shorter. In this large panorama, three layers of evolution can be observed which overlap:

— the time for tools;
— that for machines;
— and then that for systems which we are currently dealing with.

What could be the following one? We just are able to perceive tendencies.

* Tool

The concept of tools (and its complement: the instrument) emerges from developed species: monkeys, otters, some birds spontaneously feel the need to use a stick, a stone or a simple twig in order to crack a nut or reach an ant, but it does not proceed much farther. They never benefited the necessary spark of intelligence in analyzing the cause to effect relationship between the strike and the nut which gets broken. They never imagined the reproducibility of the experiment. Understanding that a bolt and a screw are made in order to be assembled provides the first step of intelligence: this could be taught to a monkey, but if it happens that the bolt has been changed, the monkey would never understand anymore why it does not work any longer.

In short, the Man became a Physicist when learning to use a stick! The tool will be considered as an extension of his hand which will improve the strength, the precision, the efficiency of the gesture and, by himself, he was conscious of that … unconsciously!

However, human thinking has been able to go farther, anticipating the actions, improving the tools, remembering the experiment gained in the implementation, imagining the arrow's trajectory, inventing a sharpener to improve the cutting edge of the axe … etc. In short, the Physicist was born!

Was it the basic need in food which triggered the intellectual process or the contrary? Does not matter. However, these mutations get printed in our brain and have been transmitted along generations. In those old times of great barest necessities, the intrinsic value of the tools, their preciousness, was very high. This made possible to create wonderful and huge marvels such as Egyptian pyramids or Greek temples obtained with rudimentary tools.

Tools, then, were synonymous of survival and one took great care of them; learning and knowledge transmission were essential. Step by step, things have evolved towards an abundance and larger ease of manufacturing, thus gradually lowering the intrinsic value of the objects. Today, we "consume" any kind of tools without mercy. Then, machine came and took over, consequently manual works got depreciated because of lesser amounts of manual involvement.

* The machine

Then came the big revolution of the XIXth century; Man who was initially limited to his muscular forces (or that of animals) discovered Energy which would to considerably extend his capacities.

Machines will get mobile, disposable, and adjustable to any use; they are substitutes to the human force in order to alleviate the drudgery of work. One is no longer attached to a waterfall neither dependent on the wind.[6]

Machines are complex assemblies of mechanical parts which are manufactured or fabricated by other machines: the machine becomes self-generated, sometimes using regulation mechanisms (automation equipment) which make functioning easier, thus freeing man from tedious tasks, the only required man's intervention being limited to enter a reference.

The evolution of steam machine followed by combustion engine and electricity shows a continuous progress. This evolution towards an increased complexity then leads, independently of man, to bring the machines together in order to

---

[6]What an irony today, after centuries of science to be dependent on the wind and wait for the miracle of the wind farms!

make coherent assemblies, no longer doing a task but more likely a "function": the systems follow next.

* The System

Tools and machines are no longer sufficient for the increasing human needs; One will get all this stuff together within more elaborated configurations using more largely automatisms, communications, and software.

Force, energy, and now logic (or intelligence) of the command are actually available. Gradually, the man's hand becomes less and less necessary; systems are more precise, more reliable, and tireless. Systems will help organizing factories, transportations, satellite communication networks, robots, …, etc.

Operating the machine sub-assemblies becomes smoothly coordinated with an efficient and an automatic operation. Using the various telecommunications means, the systems get in conjunction and interact to the point that they form complex and coherent entities (transport management or supplies, for instance) making a web extending over the whole planet.

*And further on?

This long sequence of shorter and shorter stages is currently going to be transformed into another stage we cannot clearly discern but which is logically to be realized soon. This to the point that we will say, when we will benefit the necessary distance: "Of course, it was obvious!" All of that remains the inevitable result of a patiently accumulated matter for centuries: Knowledge.

However, in this tentative projection, what we can take for sure is that, in this new foreseen landscape, Software and Artificial Intelligence will assuredly play a decisive role with also a strong implication of biology. Time has come for a symbiotic meeting of the unanimated physical world with the world of the living.

We just entered the "cloud" and we do not know yet what is behind, something big assuredly!

The emergence of present novelties

Sciences certainly get developed over the years with endless and increasing successes and this is happening, independently, in every domain. But what put the fire in was, unquestionably, the discovery in 1947 of the so-called "transistor effect" by William Shockley, John Bardeen and Walter Brattain at Bell Labs (Murray Hill).

A very few commentators pay homage to these three scientists (Nobel Prize winners) except Claude Allègre[7] (former French ministry) who had such a clear-mindedness. These invaluable discoverers are now left behind; their roles, anyway, were comparable to or even more prominent that the ones of Christopher Columbus or Gutenberg! They do have revolutionized our civilization.

---

[7] Claude Allègre, La Science est le défi du XXI siècle, Plon, 2009.

Do we have to congratulate them or slam them? The fact is, they have done it and nothing will ever be as it was before! Each of us benefits, without knowing, from billions of transistors which are working for us in our close vicinity.

Initially, this invention was only a simple bench experiment[8] which demonstrated that controlling the electrons in the bulk of semi-conducting materials was feasible. The discovery was new in the principle but had nothing to do, at that moment, with an object of practical use. It was just serendipity, the experimental verification of a far-fetched idea in a field which was not on topical issues. It took seven years of intense arts and crafts to master the Germanium material to obtain the three legged transistors we have known about at the beginning, and which were to replace the old vacuum radio tubes. Nobody, at that time, could imagine the scope of this discovery, not even the most bubbling forecasters. This goes to show that one may be wrong because of an excess or a lack of imagination as well.

We had to wait until the historic day of October 18th 1954 when the first radio receiver equipped with a three legged transistor (Regency TR1) was put up for public sale and October 1st 1959 for the first Electronic course to be given in a French university. The first book dealing with the subject had just appeared before and the word was a new one: it was said formerly, in French: "TSF" as "Transmission Sans Fil",

---

[8]— Beautiful physics, at a human scale, in the pure tradition of the Volta's cell or the first electricity driven motor.

because radio, at that time, was almost the only application of electronics.

I was lucky enough to enjoy this adventure from the beginning to the end. As any conquest, this one makes history and posterity will be the judge to decide if it was for good or evil, but what is sure is that it is no longer reversible.

In the year 2002, 10 million transistors were fabricated each month for any human (on average because some get more, others nothing!). I do not have the present data. The cost of manufacturing a single transistor is reportedly less than a single letter printed in the *Wall Street Journal*.

The first Integrated Circuit (IC) was implemented by Jack Kilby at Texas Instruments, in 1959, on a Silicium substrate and, then, the very power of the discovery appeared in its genuine dimension; there was no limit for the integration of multiple components on the substrate and the idea of "function" replaced that of individual components and wired circuits.

Currently, millions of sub-microscopic transistors are packed on Silicium "chips". Electronic functions get an unthinkable complexity, a total reliability, and at a minimal cost.[9] We should refer further to the details of this fabrication technology which were subject to extend towards domains in which they were not expected to apply, based on the new technical means which had been developed on this occasion.

---

[9]The so famous Moore Law anticipates a doubling of the performances of the ICs every 18 months. This law is regularly questioned but still remains valid up to now undeterred.

IC applications are of a universal order, ranging from the microprocessors in the computers to the chip in the credit card or even the unit in the washing machine. But another class of applications is to be mentioned that will operate as a second stage of the instrumentation rocket: I mean the new "intelligent" instruments which are to energetically "boost" the development of other sciences which were left, so far, on their own, in some unforeseeable domains.

For the first time, a convergence or even a "complicity" is to be observed between sciences; the walls gradually fall which, up to now, were separating "hard" and abstract sciences (mathematics, physics, chemistry) from "soft" ones (life and observation sciences), under the pressure of software and instrumentation. Mathematicians themselves, however esoteric in their abstract world, no more show contempt to computer calculations, they in fact are looking for them.

Another important point, along with the transistor, is the LASER[10] the principle of which was discovered by Theodore Maiman in 1960 using a ruby crystal. One says that, with laser, light became "coherent" which is not a natural behavior. It really must be said that God behaved in a somewhat disordered manner with light!

This extraordinary property makes it possible to propagate light in a more disciplined way; it can be operated in a continuous mode as well as an impulsive one, displaying sometimes enormous power within a large spectrum of wavelengths from the deep UV to the far infra-red.

---

[10]LASER for *Light Amplification by Stimulated Emission of Radiations.*

Shortening the laser pulses even leads to single photon emissions, thus leading to very unconventional properties which are still not fully explored. The photon is at the same time everywhere and nowhere! And we did not succeed putting it in a box which could lead to extraordinary prospects in the software field!

Concerning the Laser's domain, the evolution was quite fast, stable and surprising. It was a common saying, in the early days, that "the laser is the elegant solution to problems which do not exist yet". Today, however, the laser participates in lots of important applications: medical, industrial, military, spatial etc... Laser takes many diversified shapes from huge machines which stimulate nuclear reactions to microscopic components in the ICs or repeaters for optical fibers. Laser is a necessary complement to modern instrumentation.

No doubt, transistor and laser are at the origin of the science explosion.

More generally speaking, one evokes this science assembly as GNR for Genetics, Nanotechnologies, Robots (in French, we use the acronym NBIC which includes Cognitive Sciences). In the next chapter, we shall detail these points and their respective progresses.

Chapter 2

Nanotechnologies

Nanotechnology is a subject that is complex, multiple, moving, recent and still a little fuzzy. It is a typical example of the convergence of several scientific disciplines: physics (including electronics, optics, quantum mechanics, robotics, magnetism, etc…), chemistry (mineral, molecular, organic, structural, etc…), biology (cellular, genomic, etc…) and even medicine.

Technologies from the micro-electronics

Here, we are also bound to speak of the transistor because it is at the very origin of the nanotechnologies!

With the Integrated Circuits (ICs), we had to "engrave", as to say, the Silicium wafer with transistors in order to achieve logical circuits (microprocessors) or memories (USB keys). In the beginning, monolithic technologies were rather crude, using photosensitive resins, masks and corresponding basic photolithography techniques. Then a sharper precision was required for a better resolution; one used to get to UV light, X-rays, ion beams, ion implantation to finally achieve a standard line detail of 22 nm, and 14 nm in 2014. Recently, IBM achieved "bits" with the support of 12 atoms only! We are very close to the Q-bits which we will be looking at soon. One can imagine actually piled up circuits, thus gaining a third dimension to compact many more functions and shortening the transit times of the information.

The direct consequence of a better control of the circuit drawing was a much higher integration density with lower individual consumption. We are about to reach the quantum limits of atoms and quasi unique electron.

But such mastering of the "machining" of Silicium leads to the idea of fabricating very small mechanics based on the model of those we already know in our macroscopic world: gears, sprockets, racks, electrostatic motors, mobile micro-mirrors, even ... nano-cars! Such Micro Electro-Mechanical Systems (MEMS) and ICs can be produced at a large scale, as well as the ICs. They even could be assembled to give nanorobots or artificial "molecule-machines".[1]

Resulting applications are largely diversified and are part of our daily life: inkjet printer injectors, micro-mirrors for video-projector, accelerometers for play-station, or iPhone, but also in the medical or biological domain: "lab on a chip". We remain, anyway, in the micro scale not yet the nano one.

---

[1] Collection Le magazine scientifique (Université P Sabatier), No. 16.

But new and more promising horizons appear for electronic techniques: molecular transistors, spintronics, quantum calculations, photonic memories and so on.

The future promises new ongoing possibilities, it appears feasible to directly interface electronic circuits with biological molecules or even biological cells such as neurons.

All of that goes with the always increasing performances of electronics instruments: all-purpose computers, digital imaging, displays, sensors, intelligent robots, near-field microscopes, etc. New instruments that make it possible to build new components which generate new instruments and the loop get closed, seamlessly.

But as the instruments start performing better, we enter the world of smaller dimensions, the nanometer[2] scale which means small molecule scale.

This world, close to the atom, is a dark one. Light (photons), which we consider so precious, can no longer provide us with images; photons are "too big" to describe the details of the objects. The best optical microscopes are not able to "see" objects smaller than half a micron and still with extreme caution.

The electronic microscope comes to help but with evident drawbacks for the living matter. Vacuum and electron bombardment are to destroy these fragile targets.

But new instruments have appeared — the Scanning "Tunnel effect" Microscope (STM) of Binnig and Röhrer, the

---

[2]One nanometer is for a millionth of a meter.

Atomic Force Microscope (AFM) of Calvin Quate and finally the Scanning Near-field Optical Microscope (SNOM).[3] The emergence of such instruments have given rise to a flourishing series of versions combining fluorescence, near-field, space modulation, phase detection, and so on, providing a "vision" of the nano-world but also means for local actions. At this scale the photon in a way behaves as a local "electromagnetic storm" capable of inducing physical effects. "Molecular surgery" might intervene and lead to atomic individual constructions for a new chemistry.

Direct applications

Then, circuits became minuscule but at the same time "intelligent" too. A chip has been made, the size of a pin head which contains an energy captor, a battery, a computer, a pressure sensitive captor, a radio transceiver. It is of course programmable from the outside and is intended to be implanted in an eye in order to continuously measure the internal ocular pressure.

Such components are currently used that are implanted in the brain in order to electrically offset serious troubles such as Obsessive Compulsive Disorders (OCD), Alzheimer, Parkinson ... This is certainly not curative but the electrical stimulation provides with an immediate and evident relief.

[3]Jean-Pierre Fillard, *Near Field Optics and Nanoscopy*, World Scientific, Singapore, 1996.

The famous review journal Nature recently reported that a quadriplegic woman aged 58, received a digital brain implant in her cortex. She is now able, with her simple thought, to feed by herself and lead a machine to do essential gestures. In this case, the man/machine interface becomes real.

Here too, it works and it is only a very beginning. Why not imagine that brain could be "disconnected" from a body which is no longer needed and which could be efficiently replaced by a more performing machine. Software is in a permanent upheaval: It is forecast[4] that, in 2020, we could get in a Smartphone a computing power equivalent to that of the present super-computers.

In the same vein,[5] an artificial micro sensor array was implanted behind the retina of a blind person in order to stimulate the optical nerve following an electronic image captured by an external mini-camera. It works too! However, some people get disappointed that this first image is only in black and white, and is poorly resolving (only 60 pixels, but in the meantime much better performances are to be obtained[6], so the feasibility stage is already reached). However, concerning a person who has been blind for years, it is quite a miracle! It remains also thinkable to perform the same operation with blind babies and "teach them to see" while they are young enough.

---

[4]Nina Easton, *Fortune's guide to the future, Fortune,* January 16, 2012.

[5]May 4, 2011, Institut de la Vision — Hôpital des quinze Vingt, Paris.

[6]New devices with 1000 pixels are already tested in Germany.

On the other hand, we are here "in the plane of Ader" it certainly will not require 80 years for us to fly a Concorde! The day when the connection with the optical nerve will be satisfactorily achieved all the problems related to sight will get a solution. From that point, imagining an eye sensitive to infra-red or equipped with a zoom or receiving directly the TV or virtual images to get into the virtual reality… there are no limits!

Would we need an artificial eye, an intelligent ocular prosthesis, more powerful, more versatile than our present biological eye? Why not, considering the speed at which things are changing?

But to come back to the nano-world, the implications of the control of the "tiny things" are to get an overwhelming importance.

Nanometric extension

The technical mastership of the "infinitely small" world advances at a crazy pace. One has again to be conscious of the nanometric scale, that is to say a little group of atoms, a small molecule, which is very far from our human scale. None of our familiar objects belongs to the nanometric scale; only gases contain such "free" particles. Even the finest dust particles (some micron wide) are larger. We shall come back further on the possible dangers of such minuscule manufactured particles.

A new chemistry was to appear (was it chemistry or atomic physics?) the laws of which would be different from the usual ones at a macroscopic scale, thus generating new materials which do not exist in the nature and whose properties are revolutionary: similar to Carbon nanotubes, stronger than steel, or "fullerene", or "graphene" invented in 2011 at IBM.

For instance, graphene can be used to make membranes to desalinate sea water as well as to make electronic components or perfectly conducting tapes at room temperature. The list is endless!

Corresponding physics is special because we are in the Quantum domain where matter behaves as a wave. For instance, a nanometric wire does not conduct electricity in a continuous and disorderly way, but with bunches of electrons; a nanotube conducts heat through quantified vibration modes, and so on. All of that promises a wealth of applications which are still unthinkable in the chemistry domain but also in biology, pharmacy or medicine.

At such a scale, our imagination has to switch with anthropomorphism, the objects to be fabricated and used will adopt a different shape for a different utilization even if sometime we endeavor to use a vocabulary of "macro" analogies. Other phenomena also appear with which we are not accustomed, it is a new world as if we are transported to Mars!

We are in an intermediate state between the atom or the molecule and the collective state. One actually knows how to generate such assemblies: carbon nanotubes, atom aggregates, monolayers or atom islands etc.

New phenomena arise which could lead to unexpected applications:

- Quantum size effects in the aggregates which behave as quantum boxes where electrons can be confined in small numbers. Their transport obeys new laws to be discovered, depending on the make-up of the box which behaves as a genuine "super-atom".
- Exceptional mechanical properties of the spherical (carbon or semiconductors) or linear (thread) structures.
- Atom long distance transportation by interatomic tunnel effect on interstitial sites (some enthusiastic people so far are to speak of matter "teleportation").
- Virtual photon transfer between atoms or molecules giving rise to possible optical nano-communications.
- Computer modeling of new chemical reactions.

The list is not exhaustive and could lead to diversified applications: nano-electronics will push the circuits to their utter limit of miniaturization; composite materials will reach surprising properties; new types of electronic components will appear (Qbits) to fit with quantum computers the algorithms of which are to be reformulated; encapsulation of specific molecules in nano-wrapping leading to a new medicine etc.

The fabrication of a unique nanometric sized object, free or not, would be of no use; we would not know how to grasp it nor what to do with. It is Gulliver in the Lilliput kingdom! Such an object would only be usable in large numbers. It will then be necessary to develop large-scale duplication techniques in order to make them in appreciable amounts, and, from there, problems arise.

Despite it, nanoparticles (first generation) are already largely used in lots of familiar products (cosmetics, food packaging, and paints) because of their property to penetrate into cellular tissues.

The domain of Quantum electronics

Present software electronics exclusively uses a binary logic that is to say a switch between only two different possibilities (Yes or No) represented by an electric gate (transistor) which can only be open or shut. However, other more complex configurations can be emphasized such as that provided by "quantum systems", that is, atomic scale structures (nanometric).

At a so reduced scale, physics reality no longer stays as in the macroscopic human world. We enter a universe of uncertainty where the electron no longer is taken as a particle but more likely as a "wave packet" — the position of which is represented through a "probability cloud": it is here or there or may be both as well, in the same time. Then such a "door" is no longer binary, which leads to a logic richer in possibilities. One does not talk any longer of software "bits" but rather "Q-bits"!

Richard Feynman (former nuclear physics Nobel Prize winner) once imagined the fantastic properties of such system but without suggesting a precise physical realization. Today, however, the idea of an atomic scale computer is emerging again. One of the major disadvantages to overcome relies in the atomic thermal disordered agitation that cannot be avoided thus making it a major disability compelling to work at very low temperature, close to absolute zero. Some call it "cold" software.

IBM and Google nevertheless entered in the game (as well as Microsoft and some US universities) because the issue at stake is especially attractive. At IBM, the selected approach was that of "quantum wells" elaborated from groups of isolated atoms ("quantum dots") whereas Google engaged in the solution of "Josephson diodes". Others are turning to different solutions such as those promised by the light properties.

All of that still remain very prospective even if a company (D-Wave) already proposes a machine that nobody knows exactly what it is. Both Google and NASA bought this highly priced machine. It becomes obvious that if, effectively, such a system would work and if one knows how to program it properly, then the new development outlooks for the software would constitute a true leap forward.[7]

The dangers of nanoparticles

Nanoparticles are made of atomic aggregates, the size of a small molecule (some atoms), that is to say thousand times

[7] Quantum Leap, Lev Grossman, *Time*, 17 February 2014.

smaller than the finest dust particles of the urban pollution arising from combustion (coal, fuel, wood, industrial smoke, etc.).[8] They, then, cannot be eliminated by conventional masks.

From a sanitary point of view, these particles, at the limit of molecular chemistry, constitute a real gas. They are recognized as potentially harmful against cellular tissues, leading to oxidizing stress and necrosis. They are easily absorbed by the mitochondria's and the cell's nucleus, thus giving rise to DNA cellular mutations.

People are especially looking for well-mastered replication techniques in order to prevent any emission in the open which could cause adverse effects.

One also keeps seeking for reliable means to be able to efficiently destroy these particles and instantaneously stop the duplication process thus preventing a loss of control and unexpected self-maintained mutations of cells when they emulate the living. These necessary precautions generated political interpretations denouncing the possibility of a new kind of totalitarism.

Some even imagined cataclysmic[9] processes induced by a "grey jelly" that has become autonomous and invading. Others replace these discussions in the "trans-humanism" framework. Science fiction is not so far!

---

[8]To be noted: nuclear plants only emit harmless water vapor in the atmosphere.

[9]Julien Collin, Le silence des nanos, www.les-renseignements-genereux.org.

It still remains that, in the public mind, the prefix "nano" is not (yet) firmly associated to a danger but more likely something that is scientifically gratifying and commercially attractive with incentives. This could be the reason why it is so frequently and often abusively found in literature.

Genetics, Nanotechnologies and Robotics (GNR) in general

GNR in their global meaning reach various fields we shall deal with later in detail. Implications of GNR for humans today are rather large, be it in health, life time, society, or religion, all are to be considered, including the more fundamental certainties. The "self", so treasured by philosophers, so intimate, so moving and yet so poorly defined, is to be considered, scrutinized until it will become rationally "framed".

The brain, so badly known up to now, begins to be investigated with proper tools in order that it will become possible to distinguish its weaknesses and correct them before they become pathological. Biochemistry gets more precise, one tries to make bearable to the living neuron the electrodes connecting to the external world. Electromagnetic "waves" emitted by the brain in its continuous activity are analyzed by computers in order to understand how "thinking" behaves and why it is acting so.

We would like to be able to cure mental diseases in which mechanism and evolution are hardly discernible such as Alzheimer, diabetes, atherosclerosis and evidently cancer.

We would like to be able to modify the genome, correct the nature errors, or even, as Pygmalion, attempt creating perfect beings!

Machine Darwinism

This new overview of GNR will have to handle the unescapable evolution laws of Darwin because they too are to be obeyed by the machines as well as the living!

In the old time of the "neolithic computer", I mean 40 years ago, our home PCs were far from the present ones. Nobody then, even the most marijuana smoked hippy would not have been able, in his Californian dreams, to imagine the future we are living with today.

In this remote time the computer was a huge machine, noisy and calorigenic. Its "genetic code" was expressed through a bunch of perforated cards piled up in batches; the language was Fortran or Cobol. Obviously, nobody would get to the curious idea of bringing it home; and to do what?

At that time in 1968, Intel was not yet "Inside"! It was even not a start-up but a simple small company dedicated to the fabrication of copper circuits printed on a resin board, Electronic companies, like Fairchild, were busy with soldering three legged transistors on the boards in order to assemble circuits. It was just the moment when a new gene was introduced in the technology by the physicists in the form of the Field Effect Transistor (or MOSFET[10] and later MESFET)

---

[10]Metal–Oxide–Semiconductor Field-Effect Transistor or Metal–Semiconductor Field Effect Transistor.

which no longer uses the bulk of the chip but instead the surface of the semiconductor material. The monolithic technology was born.

Then it became feasible to produce, at the same time, lots of circuits in the same series of lithographic operations. The IC was there and it will evolve rapidly under the environmental pressure of many manufacturers (Texas Instruments, Motorola, Fairchild, etc. ...). The first application was a 4-bit logical programmer intended to control tool machine or even automatic irrigation systems for agriculture.

But a very astute engineer at Intel understood that it was possible to extend to a versatile, polyvalent, light, and easily movable computer. Intel refused the project and the engineer came to Paris to build the first dinosaur of the PC's family called "Micral I".[11] Programming took place directly, electrically with switches bit after bit! Heroic, but so historical!

But evolution laws immediately seized the opportunity and applied towards an endless better adaptation. A real demand existed for such a small unit; the idea of the PC[12] merged and successive more and more adapted and sophisticated versions appeared under various brands and other names.

All of that happened until Steve Jobs brought a revolution with its induced mutant, the famous "Macintosh" which, as

[11] With this Micral we were able, with Dr. Michel Castagné, to perform the first computer-driven laboratory experiences.

[12] Personal Computer.

the computer first of its kind, got a lot of new wings: mouse, white screen, pointer, icons and integrated operating system. The species definitively got a new direction towards a new race of friendly, personal, practical and household computers. The road was open to today's Windows 8, to laptops, touch tablets, iPads, etc…

Nobody, even the boldest forecaster, would have imagined some 40 years ago, such an impressive mutation neither the corresponding universe which imposed on us and which we willingly accepted. Nobody could have suspected that within such a short while the PC, a machine, would become a so friendly (?) daily fellow, a witness to our intimate life. Let us emphasize that the mutation is a continuous process, this will be reflected with the new entrants such as tablets, sophisticated Smart-Phones and mobile applications.

Natural selection, then, also holds for the machines which are to get adapted to the surroundings and the pain of going under or give rise to a new branch. So was it with steam locomotives, a mechanical marvel, a triumph of thermodynamics, which reigned over our railways during a large century and later gave way to Diesel or electrical engines that no longer spat embers! The only difference between biological and machine selection relies in the time scale which is no longer a matter of million years but only a little century or even much less nowadays.

Now the Darwinian selection rule just brought a new species: the nanometric species which will have to comply with our technical surroundings!

The Darwin's software won! But what could we expect now for the next 40 years with such incredible nanotechnologies, with electronic components so intelligent and minuscule? Would it be unconceivable to direct relation of our brains with this machine, we could no longer qualify as inanimate and the capacities of which gradually increase?

Our relationship with the PC is currently managed through an interface of mouse, keyboard (or touch screen) and display. More recently an improvement was provided by the Siri language which allows directly speaking to the Smartphone and getting a vocal answer. Would it be unthinkable (in a future to come) to consider a computer which could directly interpret our thinking and reply the same way? This would allow a fantastic symbiotic hybridization with Internet and its unfathomable source of knowledge!

Nanotechnologies and the living

Nanotechnologies not just to give rise to mineral applications they also are directly concerned with the living world.

Not to mention micro-electronics, instrumentation or nanorobots, the most impressive contribution of the technologies recently extrapolated from the Silicium processing is that of the "Lab on a chip" (or Bio-chip, or Gene-chip). This serendipity looks very simple in the basics: instead of performing biological or chemical analysis using the classical and laborious methods with cumbersome test tubes and fluid bottles, standardized micro drops of gel or reagent are displayed on a Silicium or glass substrate the surface of which is provided with an arrangement of carved mini cavities.

Then, hundreds or even thousands of usable test points can be simultaneously or sequentially analyzed using a computer assisted machine. The deposition of the product to be tested is performed by "micro-pipetting" or "micro-printing" exactly[13] over the micro- cavities. A fantastic saving of time, cost, security, consumed products and staff is achieved. We then refer to a real laboratory at a micro-scale, a "lab on a chip", as said, which is automatically operated without the help of human intervention.

The detection of the reaction is automatically activated by fluorescence (multi-chromatic) or chemo-luminescence and the result automatically stored and displayed in a computer. Sometime a mechanical assistance can be locally provided on the spot, using integrated electronic components or MEMSs.

This technique appeared relevant in the nineties and is applied now in various domains: genome analysis, proteins, and antibodies etc... The technique enabled analyses with a capacity unexpected to date. Of course, ongoing developments accelerate to improve the performances and reliability. Works, for instance, are on the way to use this technique for an early detection of the Alzheimer disease.

A program is also held in Israel to determine how stem cells proceed towards the formation of a specific tissue. These cells are usually taken from the spinal cord and must be

---

[13]This technology is very similar to that of the inkjet printers.

screened for one over hundred thousand times. Bio-chips are remarkably adapted to such a screening.

Other ideas are still on the way as for instance:

— The introduction of infra-red sensitive nanoparticles in a tumor which can be locally burned with a laser (photo thermal effect).
— Introduction of magnet particles in the very core of the cells in order to be able to move them as magnetic robots.
— Fabrication of nanometric scale devices which emulates biological systems (Biomimetic). Inversely, we try to understand the behavior of bacteria to elaborate molecular nano-machines. One looks yet to duplicate the propulsion system of bacteria which wave flagella as propellers thus providing the cell with a means so as to move very quickly.
— Object handling with optical near field microscopes.
— Assembling biological machines from DNA molecules activated by proteins: there are called "nuclear acid robots" or, more colloquially "nubots".
— Assistance to reconstruction of biological tissues by fixing nanorobots "on the back" of white blood cells which leads them on the site of the injury.
— Nanoparticles are often simple molecules that are used as a marker in a cell (fluorescence ultra-microscopy).

Chapter 3

Software and Information

Software science was born after WWII. At that time, the ENIAC computer contributed largely to the development of the first atomic bombs. Nevertheless, this primitive computing machine was not yet a real "computer" as we currently know. It was only concerned with the four arithmetic operations (with large numbers however) but was largely faster than a human could have done with a pencil and a sheet of paper! The machine performed ahead of humans in intellectual work!

The machine occupied a whole building which had to be vigorously vented in order to evacuate the heat generated by hundreds of radio lamps. Pessimists used to say that such a diabolical engine would not be able to work because it would certainly be permanently out of duty, given so large a number of poorly reliable lamps. As a precaution, some sub-assemblies were carefully duplicated. History demonstrated that it effectively worked!

Nowadays, three domains can be distinguished in the new field of computers: First the technical means, electronics, "the hard"; then the method, the language, the organization, "the soft"; finally the content, "the data".

Developing a new science

Since these heroic beginnings, there have been rapid transitions. Computer was born with its punch cards and memories woven with electrical wires and small ferrite rings. Mathematicians did not stop asking for improved specs for more and more cumbersome calculations.

Prehistorical times are gone; the transistor has come by the billions and replaced this "bric-a-brac". It has been given a free reign in the computer world. Processing speeds get faster and faster within these multiple "cores".

The number of per second operations reaches the "petaflop"[1] level in the big machines: Fujitsu "K Super Computer" crossed the 10 petaflop threshold. Our home PCs, tablets, Smartphones contain a wealth of calculation means of which we use only a very small part. But their new user-friendliness allows the non-specialist to very simply carry out a dialogue in a perfectly transparent manner with the machine through several layers of languages and interpreters.

The machine makes continuous and indisputable efforts to come closer to the user. Our Smartphones are ready to be

[1] A flop stands for a unique operation in the central unit and one petaflop points to a million of billions of them (per second of course!).

soon equipped with Voice Analysis in order to make it possible to directly speak to the machine as if it was a friend, and it is claimed that it will answer in a clear language to give you all the required information.[2] This is boundless due to the links with Internet, Google Earth, and GPS ... an inexhaustive knowledge.

Google also offers now "Google glasses", which are made up of a transparent screen, providing us with projected images similar to the head-up display of fighter jets. The spectacle frame hosts all the electronics of a Smartphone, thus allowing GPS guiding, instant information, photo snaps, video or virtual reality displays.[3] What is only missing is to be able to directly command the system by the brain ... but it will not be long before we do.[4]

However, at the very basis, the machines are run by a binary language (0–1) following a strange algebra from an Irish mathematician in the XIXth century, George Boole, and, to say the truth, quite forgotten since. Such as, in the scientific matter, the instant great idea comes to fruition well after, when its relevance becomes evident.

This new Yes/No logic will fit very well with the transistor's specifications and will give rise to a new kind of electronics: the digital technology.

---

[2]It is already the case with SIRI. A real explosion is going on in the man/machine relationship through new gizmos.

[3]In order, for instance, to help a surgeon when operating.

[4]It is not yet known which insurance company will accept to cover the risks related to a pedestrian or a driver wearing such glasses.

How and why "digital"?

Here it becomes necessary to give some explanations, even if they could appear a bit sketchy, because most people ignore what is behind this term which they however use currently. "Digital" is the key technology in all of our machines, it controls every communication systems, it is everywhere. Even our Television itself has become digital. Some words are then necessary for such an important innovation.

So far, the "electrical signal" provided by a microphone, for instance, was an ongoing electrical voltage which varied with time following the variation of a sound or the voice. This signal was to be "analogous" to the sound and so it was also called an "analogue signal". It was said that it was an electrical "print" of the sound and thus the information could be recorded in a continuous and oscillating way on a vinyl disc or a magnetic tape, for instance. Such a signal is obviously fragile and sensitive to external perturbations if it has to be transported or stored. The corresponding induced incoherent deformations are called "noise". Everybody had sometime the opportunity during a phone call to hear such unpleasant sizzle or whistling which could make a conversation impossible. One then instinctively found ways to get rid of it by "coding" the message, for instance by a repetition of the words or by saying "two times eight" instead of 16. This operation is called "adding redundancy" to introduce a security in transmission of the message. Everybody understands that without the help of a drawing. However, a drawing will still be necessary for the following.

Things will go quite differently with the "digital" language. A little scheme is required to be able to follow the successive operations which electronics will perform automatically and sequentially.

In a first step, the "analogous signal" is chopped to give "slices" which will be "quantified", that is, rounded to discrete values following a predetermined scale. Then the "samples" are "digitalized" according to a writing code and replaced by the corresponding series of standard electrical pulses (or "bits")[5] in some way as the Morse's code does. These pulses are packed in the time interval between successive samples. Their disposition reflects the height of the sample. The extraordinary speed of the transistor switching allows the performance.

The signal which is to be transported or stored does not represent the voltage itself but a measure of it. The main benefit of this strange procedure is that, upon arrival, the only thing to do is to check if there is a bit (1) or nothing (0) at the "top" of the synchronization clock.

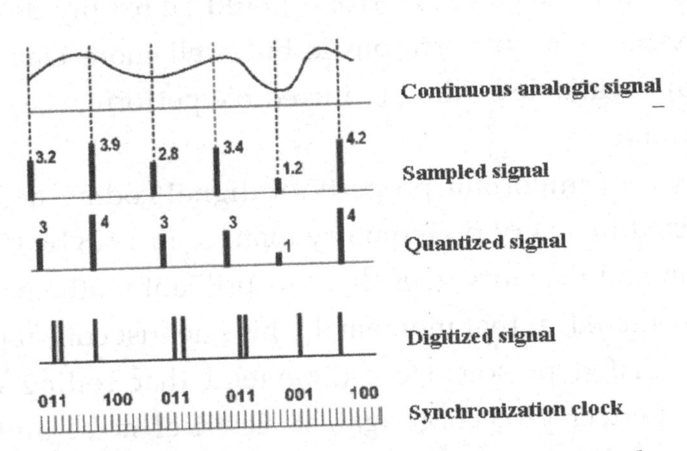

Transformation of an analogous signal into a digital one

---

[5] "Bit" is an abbreviation for "Binary Digit".

The initial signal will be later reconstructed by reproducing the same operations in the reverse way. We may need to arrange the reconstitution in such a way that signal distortions can be neglected. For more convenience, the bits are usually grouped into "bytes" (8 bits).

Every information, numbers, letters, control signals can be written in this conventional form. These bytes can be piled up by billions on hard discs, DVDs, USB keys and will occupy a reduced place. Such writing also benefits a valuable asset: it is durable. Unless we tear the disc into pieces, the information should survive over time and wear, intact.

If the surrounding environment was to be especially noisy (distant transmissions), it could be even possible to change the bit series of the message using conventional codes: for example, repeating each bit three times; that is quite easy and, at the arrival, the possible errors should be readily detected. This example is quite simplistic but well more elaborated codes are available leading to incredible performances in the noise immunity.

This very important property of digital coding has been recognized in a very premonitory manner in 1948 by Claude Shannon and the story goes that this brilliant mathematician was considered, at that moment, by his scientist colleagues as an illuminated person. He did pretend that coding could bring a perfect protection against electrical noise and that there were no fundamental limits to telecommunications. Still a visionary struggling with disbelievers!

It still remained that the predictions were valid and because of Claude Shannon we dared to send probes to the remote space as the two "Voyagers" which just came out of the Solar System,[6] some 20 billion kilometers from earth. We keep on accurately receiving the images, after a 35 year travel, in spite of the distance and their low power emitters. It is because of Claude Shannon that the robot "Curiosity" landed (or would I say "marsed") on planet Mars, regularly obeying and sending the results of its analysis in the search for possible traces of life on this planet.

Of course, "digital" is not only a matter of space and large distance communications; it is useful on earth too! One can find it in our computers, our medical imaging systems, our TV sets, our essential smart cards etc. ... Everything is to be translated in "digital" today.

Software

Then, it took the computers to learn how to work. It has been necessary to provide them with an (or some) internal language(s) in order the machine to manage itself and communicate with the outside world. These languages, along with the information they are conveying, constitute what is called "software" which equip every computer. Without software, a computer is a dead machine.

Every performance subtleties of the computer rely on integrated or external softwares. Such "programs", at the beginning,

---

[6] Stephen J. Pyne, *Voyager*, Vikings Ed., 2010.

were hand written, line by line; step by step; so they became big and complex and required coordinated teams of specialists ... and also powerful computers. One actually can begin to say "computers are made by computers"! We are nearly crossing a threshold towards the autonomy of the machines.

It is still the Man's responsibility to define tasks and the way to run them. A software is generally specialized in specific and defined applications: office automation equipment has nothing in common with the software that guides a robot. Nevertheless, a general tendency arises for a larger range of application, for more refined requirements, for more opened interoperability and all of that leads to larger and larger size and complexity of these tools. This consequently induces flaws or inconsistencies (the bugs) which are to affect the functioning of the machine.

The increasing performances and the decreasing cost of the components along with the evolution of multicore processor pushes in the direction of larger and more sophisticated softwares to that point that some call them "obese".

A difficulty which often arises in the conception of softwares comes from their required adaptability to new and sometime unexpected situations. We talk, then, about agile software engineering, parallel structures etc. ... Such new approaches are especially required for Artificial Intelligence (AI) processes and even more for human simulation (we will deal with that point in the next chapter in more detail). However, a fundamental difficulty lies in the obligation that we have to make the programing in an explicit formulation.

Current advanced software (including biochemical simulations, product monitoring, defense systems...) involve tenth of million lines of codes. The complexity and the need for optimization are such that software no longer remains at a human size. Specific control and development software are required (IBM). Such complexity levels are already higher to that required in a human brain simulation! Models would be available as soon as we achieve the identification of the involved biological mechanisms. The making of a copy of the brain would be straightforward.

Information and communication

We are faced with a new "matter", intangible but so dense and today ubiquitous: Information. Magic and vital fluid of the modern times, information lays in any place; it has become as necessary as the air we are breathing. It travels fast and everywhere. One can no longer live without being instantly informed. Hence, this philosopher illusion has transformed into a physicist's substance.

Writing and later printing allowed information to get in a stable and transferable shape. The Encyclopedia formerly gathered the whole knowledge in the world, but, here too, the timeliness is explosive. Two decades ago, as the researcher, I had to go and search for information in many reviews at the Library of the University. I had to laboriously flip through the pages in order to find an interesting paper which was already six months or a year old, the necessary task to the preparation, edition and distribution of the review which often came from faraway. Hopefully, we had the opportunity to make photocopies which helped to keep a trace and work more easily. However, the process was awfully lengthy, tiring, and tedious.

Today it is quite different; information reaches the computer in a jiffy, it is selected, organized, ready to use and is available in abundance. This fulgurance largely participates in accelerating the research work. The only remaining problem will be… to struggle among too much information!

Of course, the question can be asked: "what, the heck, is information?" Does a poem of Oscar Wilde really contain information? May a simple noise contain information? Must we associate information with a concrete, tangible, precise meaning?

The physicist gets around the question without a direct answer.

He approaches information under the unexpected angle of the probabilities. In this way, the exact meaning of the information, as it stands, is of no use; the only important thing is to be able to allocate it a measurement, a price, a numerical value. Probabilities offer such a possibility; the evaluation could be *"a priori"* if nothing is known before or in a significant manner if one already disposes of prior knowledge. Obviously, the more this knowledge is, the more the remaining part of the information to transmit is reduced and its value too.

Information is related to its probability to occur, then to the ignorance we have of it. For instance, we pay for a newspaper of the day because we are ignorant of the content; the newspaper of last week is of no informative value. On the contrary, how much should we accept to pay for the newspaper of next week which should contain, among others, the future stock exchange values? Information is more valuable as it is out of reach.

But the value of the information is also fugitive: it does last only till the time of the communication, a moment after it loses any value, it becomes acquit. But it has nevertheless to be stored in case it would be worth transmitting to another ignorant person!

Without entering the details, this measurement of the information value, this price, is expressed in a special unit: the Bit[7] (not to be confused with the electronic impulse which bears the same name). So are evaluated the rates of information in the communication channels which actually relate everybody to everybody on the planet.

This information will be "compressed", stored in various electronic media, and distributed to whoever it may concern. In the last decade, information revolution brought us Internet, Google, YouTube, Facebook and various other social networks where everybody can freely (?) chat and share documents.

Currently, the New York Times releases more information in a week than all information received by an 18th century philosopher during his entire lifetime. Our knowledge

[7] Bit is here for "Binary Unit".

doubles every 2 years, in 2020 they will double every 3 days. About 40 exaoctets (320 billion of billion bits) of new information were produced in 2010; such a volume exceeds all that have been produced during the previous five millenaries! Is this not an indication of a frightening acceleration? To what point such frenzy will still be humanly bearable? How to find one's way under such an avalanche? Will our brains be up at the task ahead?

The Web is evidently the conducting media and the essential storage basis of the world wide information. These information, these data, are permanently stored, circulated and auto generated between enormous server bases equipped with thousands of big hard drives and distributed over the whole world: 2.9 million in the so-called server farms in the US alone.

This accumulation is astronomic and in a constant growth to such a point that a congestion[8] is surely to happen. Software dedicated to "search and find" documents gave birth to a professional branch for very thorough specialists: the "data crunchers". This was recognized a necessity as well by the storage companies such as the giant Google or by smaller industries or businesses which manage stocks, archives, documents more and more invasively. The "cloud" is going to become the center of all information.

In the spy/military domain, Europe recently raised concerns of the invading activities of the NSA which collects and stores

---

[8] 90% of the data in the world were generated over the past two years.

with impunity all the stuff circulating on the waves (this, obviously, is not in any case a scoop, this activity was known for quite some time without being on the front page of newspapers).

Internet knows nothing but can tell you everything

This chapter could not be closed without dealing with the unavoidable information vehicle known as Internet. As a matter of fact, Internet is in no way intelligent, it is only a kind of directory, an enormous memory diluted in a huge communication net. That is enough for a search engine to find out the closer references to the proposed keywords. This is only a referral, the ultimate decision, the choice of the relevance, the final appreciation remains the preserved domain of the partner; he only knows what he is looking for. Internet cannot answer a question (at the moment) and that is what Google is struggling for by injecting more and more AI in order to make Internet a real partner, beyond the primitive state of the keyword.

In spite of the present shortcomings, Google currently stirs six billion of search requests, four billion videos, 500 million mail boxes, and that, every day! One cannot any longer live without Google[9]!

This is all the more the case as Internet is more and more to be used from a phone (4G), that is to say, with a speech input. The future is tending towards the elaboration of an advanced language which could allow back and forth

---

[9] Une semaine sans Google, Martin Somat, 01Net, No. 74, 2014.

conversational exchanges (chatterbot) similar to what could be a human conversation. The machine should humanize when becoming intelligent and meeting the Turing test.[10]

The memory space proposed by the Web is so widespread and the access already so easy that it is proposed to alleviate the charge of the operating of the PCs by directly working on line with software on the Internet 2.0. One no longer needs to install cumbersome software like Microsoft Office which will be accessible for free online; the geeks call this new workspace: "the cloud". Such way of doing business is to be widely used because it is cheaper, and easy to use; everything is done by remote access; even industrial processes are to use this technique.

All our life is then associated to this spider's web: banks, transport, industries, medicine, services, planes, armies ..., everything. Hopefully, the delocation of the facilities helps to guard from a local catastrophe. Caliph Omar may put the fire to the Alexandria library; copies of the data will still remain everywhere!

With Internet, advertisers, with the blessing of Google, got into the habit, when we "surf" on the Net, to intercept all of our procedures, to identify our partner, to draw an image of us, the "profile" of a potential customer; all of that is performed automatically and without any disclosure. They know our purchases, our providers, our personal preferences and so they are able to adapt to our personality the advertisements

[10]Turing test seems to have been recently satisfied.

they launch on the Web[11] pages. It is rather difficult to get rid of the small spies (cookies) they put on our PCs and which soon come back if we succeed in removing them.

Concerning the University domain, free courses are already proposed on the Net in order to better share the knowledge, but three famous US universities[12] have done better. These universities used to collect expensive registration or tuition fees and this benefitted for a real financial autonomy. These Universities just decided to put their courses on the Net for a free access (MOOC[13] programs). Corresponding degrees are conveniently offered to people who satisfy the organized exams for free. On top of that it is quickly observed that these teaching programs are quite serious and of the highest level in the most specialized sciences. Then it comes as evidence in a first glance that such a politic should be suicidal, but ... it actually is not.

In fact, such Universities are able to recruit their students at a global scale; they are also able to select the best ones, have them formed at the best specialties ... and sell them to the companies (preferentially American) which are looking for

---

[11] The "pop-up".

[12] This applies to Stanford, Harvard, MIT. The others (Berkeley, Penn State, Caltech, etc. ...) are already running to catch up.

[13] Massive Open Online Courses: following the US initiative first MOOCs already appeared in France thus generating quite an intellectual storm, so challenging the all educative system we call, here, colloquially "the mammoth". See: Emmanuel Davidenkoff, Le tsunami numérique, Stock Ed., 2014.

them and are ready to pay for high level collaborators. It becomes a real business of human merchandise! All of that leads also to a US monopoly and formatting.

Sebastien Thrun (Stanford Univ.) tells us: "within 50 years there will only remain 10 universities in the world". He would have added that they should all be American! But he might be quite pessimistic; this will likely happen in a closer future.

Hackers and Co.

The Web is already a playground for institutional or commercial intruders that are more or less easy to identify. But on the other hand a hardly identifiable plague appeared which was rather foreseeable: the hackers. Very similar to the Barbaresque pirates of the old times, they set up on the Web as predatory spiders!

They too are looking for information or even they try keeping it for sharing with readers. It will lead to an endless game of "cats and mice", and God knows where this will lead us. WikiLeaks puts the all sensitive secrets on the public stage, where recently some irresponsible individuals displayed on the Net the true DNA sequence of the smallpox virus!

Professional "spy-pirates" (industrial, politics, journalists, secret services, terrorists, etc. ...) consider that the Web is the battle field of modern times; one can imagine that, one day, nasty surprises will merge. Anyway, the invasive globalization of information, the publicity of our privacy (already accepted by so many people) brings us closer to a life where Big Brother will be the King.

It can be added that the fragility of the radio-electrical links and the dependence we are of them means that, currently, an "electronic war" should be well more destructive than a conventional one and also bloodless (at the moment), with an invisible foe. A formidable task force or battle fleet should be placed *"hors de combat"* with an electromagnetic weapon or malicious software, not to mention the vulnerability of planes or satellites.

A real nightmare is a possible cyber-attack by unidentified foes (political groups or terrorists) against public infrastructure such as transport (air or railways), the electrical grid, or, more simply, elevators which are, as every system, connected to Internet. Serious harm could arise, without suspicion, from such a coordinated assault. We all are dependent on Internet whether it is wanted or not.

But how does hacking works? Every computer or digital system is run by software which is as far as possible protected against unauthorized intrusion. Nevertheless, programming thousands of code lines is not an easy task, some unwanted errors can be done which we call "bugs". The hacker takes the opportunity of these flaws either to perturb its functioning with a toxic code, either to get an access on the site and steal protected information or install a sleeping code to wait for something and start an action later. It is quite difficult to foresee such an attack and even the dedicated anti-virus softwares have to struggle with these always evolving aggressions.

When such a loophole is identified by a hacker, he calls it a "vulnerability" or even a "zero day" because as soon as a bug is detected it can be (and must be) immediately corrected by the owner of the site. Then this gives rise to a special business; a "vulnerability" can be readily sold to the owner of the site who is happy to quickly fill the breech and is ready to pay for. There is even a lively black market of vulnerabilities but information about it is scarce.

Researching and selling this new and soft merchandise became an active business and companies have been created to deal with the matter (iDefense, TippingPoint, Vupen and others). Some vulnerabilities have been known to change hands for hundreds of thousand bucks each! They ironically dubbed: "we protect you from people like us". The preferred targets are the back-up sites because they generally are less protected. Of course, other companies specialized in countering prevent intrusions with permanently up-dated anti-virus software.

Undoubtedly Internet, below the surface, has become a war zone where criminals, law enforcement, military etc. … are playing "cyber warfare" on an international war field. NSA and FBI are fond of implanted "sleeping" surveillance software on targeted computers without them knowing about it. Budget defense amounted $5 billion for 2015 cyberspace operations and there is, up to now, no currently binding statutes that are to abide by these activities.

Now another aspect of Web trafficking is given by the so called "Deep web". We do not know for sure who is exactly at the origin of that, likely the US military. It is a cyberspace dubbed "The Onion Routing" (TOR in brief) because the back and forth messages have to cross successive encrypted layers in such a way that any identification of the origin is possible (even by people who built it), thus providing a complete anonymity that cannot be indexed even by Google. Of course, NSA, FBI and so on are desperately trying to figure out how to crack it, seemingly without any success to date.

This appears as a huge bazaar where any kind of illegal goods can be found (from drugs to Kalashnikovs). Things are unlawful. The suspected volume of exchanged information on this media could be much larger than that on conventional Web. Deep Web is a real nightmare, a haven for any traffickers, thieves and pornographers. Jihadists largely use it to communicate in a guaranteed discretion. No traces left!

On top of that this was combined with the arrival of an untraceable electronic currency: the Bitcoin which makes it possible to pay and be paid anonymously. This makes everybody happy. Bitcoin is a purely virtual currency and can be transferred from one digital wallet to the other without any bank brokering the transaction or levying a tax. Legal businesses accept Bitcoin as well as the less legal ones.

Robotics

Robots actually invade our environment, they become closer and closer to humans and they still clean up our houses! Some of them, which can be considered as "humanoids", are moving around on their two legs with a certain degree of agility. One can foresee, in a near future to hire such a "servant" which will share our life, at home!

Japanese are nowadays the champions of the humanoid replicas: the new Science and Technology Museum in Tokyo now benefits from the help of robotized android hostesses. In a very impressive manner, Kodomoroid and Otonaroid (the names of the "girls") are able to chat with the visitors and answer their questions! Some ill-intentioned gossips claim that Kodomoroid now ambitioned to present a TV emission of her own!

Beside this surprising curiosity the fact still remains that Japanese have a very serious problem with their aging population and they expect to reach a solution with robots that could be specialized in domestic help. This could help preventing the troubles generated by an inverted age pyramid.

In our today's civilization, robots and automatisms are already present everywhere, they "cleverly" perform some repetitive or precise works we no longer want to assume[14] or we do not want to pay for a human to run them. We do found robots in the factories, to assemble or paint cars, manage the warehouse, machining parts, in the laboratory to count red blood cells, in the hospital in sensitive operations (inner ears, laparoscopic surgery…) and even on planet Mars!

They are programmed for specific tasks however they also can be evolutionary, reconfigurable, and intelligent

---

[14]Jean Paul Laumond, La robotique, une récidive d'Héphaistos, Fayard, 2012.

(in some way); they are able to learn by themselves, to improve their skills by e-learning. At the moment, man keeps the lead, the surgeon would not trust too much in the machine; nevertheless he gradually accepts the help, for instance, of artificial realty simulations. Also, remote chirurgical operations become common practice, possibly from one continent to the other.

As well, any Airbus plane is piloted by robots which the pilot trusts in, even in a zero visibility landing. Of course, when an instrument fails the computer returns to hand immediately, but it is sometimes too late to react correctly and an accident occurs; as it was the case on the flight Rio–Paris some years ago.

Very recently, Google carried out the first car without a driver (the famous Google car). They cover, on their own, a distance of 500,000 km on the Nevada's roads in a dense traffic, without any car crash. The American government[15] proposed 3000 drivers to get the equipment in order to perform a large-scale real life test.

Still worst, we just interred in the "intelligent plane" era: the first automated unmanned civil flight (without passengers) just happened in England and it is already emphasized that a "free plane" service could be proposed in a near future, very similar to the urban "free bike" already proposed in

[15]France administration would never have authorized such a test. "Nay sayers" prefer to wait for a US patent! Nevertheless, Renault as well as others threw themselves in research programs for a "connected" car.

many towns. These planes will be rentable and will carry four passengers from town to town (or better airport to airport) for a purely automatic flight. It is no longer fiction science but likely tomorrow realty!

Robot is a Czech word the meaning of which is "work" or "slave" as well (which is somewhat the same). Robots are exemplary workers: they do not need to be paid, they are not union members, they are able to work 24/7 without any pause, they never go on holiday, they never are tired, make no mistake, have no higher ambition, no social requirement; a few electricity is enough for their happiness. As long as a robot takes care of a task from A to Z, the direct operating cost almost disappears.

Recently, it has been shown that robots are even able to directly share experience between them (Robot Earth network) and so improve their knowledge. They can know where they are and find consequently their way. The huge container harbor of Rotterdam is completely managed by robots: no worker, no trucks on the wharves.

In the practical field, one can mention the stables which are managed by a simple computer.[16] Cows are equipped with an identification chip in relation with a personal folder in the computer which contains all information concerning its pedigree, medical data, milk performance, etc. ... The cow freely goes to the door in order to be milked; it is identified, taken over by the machine in a dedicated box and automatically

[16]Jean-Pierre Lacan, Midi Libre, February 22, 2014.

milked. Back to the stable, the machine cleans the box (and the cow), provides the cow with the fodder. Everything is automated and the farmer, from home, can follow the operations on his smartphone and check if everything is okay for his 100 cows. In case there is any problem, the computer sends the corresponding information (sickness, weight loss, unusual behavior, etc.) Many cattle farms in France already use such robotic equipment.

Thus far, robots are only an extension of the hand, however, some of them get more or less intelligent but the ultimate decision remains the man's domain. In a near future it might not be still true and the computer will decide by itself (have written "by himself" might I). This is to the point that a special network has been to be installed, last year, especially dedicated to robots; an Internet where they are able to find (by themselves) the information they need for their autonomous robot life: Robot Earth.

But robots are not only used to paint cars or tighten bolts (or grow cows!). Current projects on the go are dealing with every domain of human activity: medical or domestic cares, accounting, or military. Non-human intelligent fighters are expected, tireless, invincible, obeying the orders, carrying heavy charges,[17] detecting the threats and so on. Various versions are already experimented in the US army laboratories

---

[17] The famous mule created for the Army valiantly carries 200 kg of materials across the most difficult rough terrain, without falling (in such extreme case it will recover by itself). The only unstated difficulty remains ... the limited performances of the required battery.

(and others) to make armed (or not) vehicles completely autonomous. Bill Clinton himself confessed that "Human is the weak link of the to-morrow wars".

The famous "drones", unmanned planes up to now controlled from the ground, are now programmed in AI to be able to fly and take their decisions by themselves to assume the mission with a pinpoint precision, without any radio link to the ground. It then becomes impossible to jam or detect them. They are fully autonomous and too bad if they aim at the wrong target; but, do not forget, it also could happen (and may be more likely) to humans as well!

Obviously, the man's intelligence, versatility, adaptability to unforeseen situations are still largely greater than the limited reaction of the robots... however, AI turns out to be more flexible, more adaptive, needless to go to Mars with curiosity, practical applications can be profitably developed which are more open, more autonomous, with more expertise, for e.g., the robot especially designed to prune the vines. It already is experimented in the famous Saint-Emilion vineyard.[18] The machine, of rather small size, is equipped with six cameras and arms carrying pruning shears; it is able to find its way by itself between the rows of vines, to evaluate each vine stock to trim the ends at a specific place as a specialized worker would do. However, this is not an easy task; it requires experience and intuition because there are neither identical stocks

[18]This is quite surprising as one is aware of the careful attention the Bordeaux's vine-growers pay to their vines!

nor branches. In addition, the robot establishes a precise bill of health for each stock and reports back in case of something wrong!

Of all the periods in history, the machine, let alone the robot, was considered as an old enemy of the worker from whom it stole the job; this remains especially true nowadays. One might obviously object that creating robots requires hiring workers and this induces the endless question of the worker's reconversion and mobility.

We have just learned that a large Chinese factory (Foxconn) which assembles electronic parts for Apple, Nokia, Samsung and others, has installed 10,000 robotized workstations on its production lines and intends to triplicate the number soon. That means that a mob of "foot soldiers" will end up on the street without a union to take care of them. The robot no more satisfies in hard and heavy duties, it also gets closer to the human in careful, delicate and diversified tasks.

There is no doubt that, the day an electronic brain will be available, even limited but similar to a human one, robots shall get a much larger importance; and this is what is to happen. This we are to discuss in the following section.

Robots versus Man

Can robots even replace Man, at least in limited and well-defined tasks? We already know many examples of (Japanese) human-like robots or four legged robots which are able to walk, get up, climb … speak! A curious recent video clip[19] is illustrative of that; it is very indicative of the adaptability for a robot to emulate a not so obvious human activity.

The aim of this is a Ping-Pong game that is arranged between a robot and a high level player. The organization of the video shoot is not so clear but it still remains that comments will be valuable to qualify both behaviors.

The robot is made of a fixed column bearing an articulated arm which holds a classical racket. The robot is positioned rather close to the table but it is not able to move neither laterally nor in depth. This is a large disability and makes a big difference with the player who is able to get acrobatic postures.

Unquestionably, the robot is a real technical performance as for the mechanics (hard) or as for the intelligence (soft). Its handling of the ball is quite astonishing, fitting for a juggler! It surely is equipped with cameras which gives it a 3-dimensional view whereas its arm allows contortions a man could hardly be able to perform (rotating wrist). The computer does its best to give it a quasi-human behavior.

The first part of the match shows an unquestionable domination of the machine in agility, game power, accuracy of the shots. The player is dominated by the machine without any problem … until he discovered the weaknesses of the robot that is to say: a short ball flush with the net or a hard smash

---

[19] http://www.youtube.com/watch?feature=youtu.be&v=tIIJME8-au8&app=desktop.

in the corners of the table. Therefore, the game was over for the robot.

What is the lesson to be drawn from this game? Seen from the robot's perspective, in spite of an evident technical prowess, it finally behaves as a "machine to return the ball", unable to generate innovation or strategic inventiveness. Seen from the player's perspective, he shows an evident inferiority in the normal game but an unquestionable talent to find the weak points; then he only has to replicate the same shot, endlessly ... as a robot.

But in the main, neither the robot nor the player should grow from the experience. It is anyway certain that the robot will benefit future improvements which are to correct its weaknesses and then to win will become well more difficult!

In another domain of robotization it could be questionable if a vocal server would be able to supplant a human. Everybody has made the experiment, trying in vain to join an insurance company (as an example) without ever getting the right answer neither reaching a human voice on the phone!

Nanorobots

Last but not least, there is a domain where dreams have difficulties to become a reality: the nanorobots. For a long time, novelists and other cartoonists have imagined vehicles, the size of which have been reduced to a microscopic scale to the point that it becomes possible for them to enter and freely move in our body, looking for bad cells to shoot or clogged arteries to clear.

Then a miraculous prefix was discovered: "nano". The step was taken: everything will be "nano" even if the dimensions actually are not. But at a nanoscale everything is "atomic" then quantified; the macroscopic continuity of our world is definitively lost. It has never been possible to shrink an atom!

Of course, the internal exploration of the body is already performed with very small probes (arteries, digestive tract, coelioscopy, etc. ...) more or less autonomous or guided, carrying sensors, cameras and so on but in no way nanoscopic. Such applications certainly do exist and will benefit improvements, but they will remain specific and definitely non-nano.

The access to the nanoworld has developed myths but it remains the world of small molecules. It is hardly conceivable to make a non-biological complex system (computer, engine, energy supply, tools...) with an only handful of atoms! The resulting robot will certainly be too much cumbersome. It should also be as unconceivable to let such a system free to move in the body without any means for getting it back.

Of course, it is still feasible using a tunnel effect microscope to build a "nanocar" where the four wheels could be made of a group of atoms mobile around its axis; one can even make it move by lightly pushing it, but, at the moment this still remains as a laboratory curiosity.

It seems to be that, till date, the only promising perspective relies on biology. One knows that it is not easy to subject a biological cell to a chemical treatment because the walls do not easily ingest the portion you want it to swallow.

On the other hand, this portion after a long journey, randomly in the body on the look out of its target, becomes polluted and degraded. It has then been emphasized to confine the medicine-molecule in a protective "jacket" which presents a preferential affinity with this particular plasma membrane. It is, then, to ascertain that the molecule will be preserved and will be accepted by this only cell. Obviously, the process still remains erratic; no guidance can simplify it. Typically, cancer cells are concerned by the process.

It can also be imagined that self-propelled bacteria by the means of their flagella could do the same work. Then there are no longer computers, mechanics or tele-transmission involved. It is more a problem of chemotherapy than robotics. The nano-combat-tank-submarine armed with cannons and bazookas only exists in the cartoons.

Nevertheless, the nanorobot race is launched. Nobody knows exactly what could result but unexpected things will happen for sure. In the US, bankers are investing in the field, which speaks for itself. The gap between the lab and the common use will be quickly filled.

AI

What is intelligence?

That is the first question to be asked before dealing with a possible software model.[20] Our spontaneous anthropocentrism

---

[20] French TV emission: "C-dans l'air", TV5 02/13/14.

could decide that only human species are gifted of this attribute. However, all things considered, looking at some animals (and some humans contrarily) there must be doubts. We shall talk further of the brain in more detail but the basic mechanism is there, even if it requires a strong correlation with the body and especially the five senses.

Intelligence might be defined as a faculty we have which allows us to figure out and adapt to unexpected circumstances by information processing with the aim to achieve anything. This supposes a large flexibility, but reflects an only aspect of intelligence. What happens to intuition, art, humor, creativity, emotion? Are they to be associated with intelligence? Is intelligence to be strictly limited to causal, logical, deductive reasoning which is so well-developed by mathematicians? What is happening of consciousness?

Ibn Khaldoun distinguished three categories of intelligence: discerning, experimental, and speculative. The essential contribution of the culture must certainly be added, that is, memory. Some, like Kolmogorov, have been trying to formulate a "mathematical model of the complexity" from training by optimization of the probability for each possible future but it would be limited to the only analytical aspect of the problem. All of that remains quite obscure!

To go farther in the interpretation of intelligence some tried to schematize by the conception of a "conscience field" (or "cognitive map", for the computer-geography scientists). When this field is kept out of the will, it is filled with elements randomly, incoherently emerging from the memory in order to be examined, evaluated and then forgotten to make room for others. They are only suggestions, calls for thought which vanish as fast as they come if efforts are not taken to keep them. Some say in a caricatured way "the chatter of the mind" (in French *"la folle du logis"*).

But consciousness is also driven by the will to appreciate an idea, an image or a sentence in a coherent conception. So we are led to the concept manipulation, the generation of "ideas" and then the elaboration of coherent logical reasoning. Intelligence is made of memory but also of a faculty which is not necessary to the computer because its memory can be indefinitely extended: oblivion.

We are then in a too much complex situation to draw a universal and exhaustive model. The machine to accommodate an AI has to satisfy with effective but piecemeal issues even if they could be sometimes very sophisticated.

We also have to remember that intelligence is not exclusively cerebral; it also involves the whole body which sends messages and calls for a response. Intelligence, to proceed, needs the help of the five senses which provide an interface with the external world. To read a music score, to kick a ball requires some kind of intelligence (may be fewer in the last case!). Reflexes are an integral part of intelligence; there is evidence of this in the fact that education is required to make them performing. All of this has to be taken into account when dealing with the development of an "intelligent" machine akin to the human.

How to make a machine intelligent?

Various approaches were proposed under different generic names; cybernetics, operational research, cognitivism, data mining, etc., based on the field and the level of performance they are required to follow. But now AI is a general appellation universally accepted. However, we still remain rather far away from the versatility, the universality, the creativity of a human brain!

A possible historical origin of the AI could be found in an anecdote assigned to James Watt about 1750: he was a young apprentice and was in charge of a steam machine the power of which was visualized with a rotating system of balls where its spacing indicated the delivered power; his role was to maneuver a throttle in order to correspondingly reduce or enhance the delivered stream of vapor. A very tedious task! One day, he got the idea to attach the lever to the rotating equipment with a string in such way that the machine by itself raised or lowered the power as required following the position of the balls. Watt's regulator was invented and that was a step further from a simple automat. To some extent, the machine looked after itself and became a little bit more intelligent. This story clearly shows that a system as complex as a man with his brain, his eyes and his hands can be profitably replaced by a simple string when the task is carefully defined and limited. In this particular example, the intelligence, of course, is closely limited to this only task.

In any case, an approach of the possible developments must follow two distinct paths:

- Equip each machine with a proper intelligence limited to a precise task in order to improve its functioning and make it more communicative, reconfigurable or evolutive. That is already done on a large scale. Such "intelligence" should certainly be limited but optimized and oriented; it would be profitably used in robotics and will be represented by dedicated softwares.

The structure of such a program will have to be a direct and functional logic, without the aim of imitating a human mechanism even if a certain amount of "fuzzy" logic or adaptability could be integrated in order to deal with non-directly provided situations. The predictable role of "Robot Earth" will certainly develop in this way.

- Assemble individual, juxtaposed, associated and coordinated blocks, somewhat as a human intelligence does. Many efforts have already been taken in this direction of simulated human intelligence but many points still remain to be solved in the human behavior and its brain organization before reaching a coherent simulation. It is also by no means clear if such an imitation is the best performing solution in the long run. One could remember that planes were not elaborated by imitating the bird's flapping-winged flight.

Presently, then, the AI is kept divided in compartments dedicated to well-defined functions; but, the results are there; one has become able to analyze information, to interpret them, to evaluate them and to restitute them in an elaborated or synthesized manner. One knows how to cleverly analyze an image and draw from it the meaningful elements for a given application; one knows how to analyze the speech, understand the meaning (may be it is too strong a word) and even construct an answer using a realistic enough vocal synthesis. Yet we stay far from a universal, clever machine.

If such a machine was to be designed in the future[21],[22] from the computers, it is certain that it will be foolproof, impersonal, extremely fast, without any blank, any emotion, any doubt. Its logic should be merciless and largely duplicable. It will be self-improvable and will surely very quickly exceed the learning performances and capabilities of the humans; this to the point that it would become, may be, uncontrollable, which could lead to a violent "singularity".

Hopefully, there still remains (for the moment) a large gap between intelligence and power in our present world. Intelligent people rarely contribute in the decisions about our destinies, alas! It all will depend on the connection (and it shall be necessary) between the computer and the external world. This connection, in the man's case, is fulfilled by the

---

[21]Proponents of the "strong intelligence" do not care about the objections, the equipment is are already available.

[22]Software WBE (*Whole Brain Emulation*).

five senses and the control of the body. If such a machine would have to exist, "we should not be in a position to negotiate, as well as a chimpanzee or a dolphin is not in a positon to negotiate with humans".[23]

To make things worse and more worrying, the "mineral" version of such a machine issued from the Silicium technology is not the only possibility to be taken into account; biological logic machines just began to be experimented; they no longer use ICs but DNA molecules assembled on purpose to perform a networking logical function; such an elementary system is already operated. Its only drawback at the moment is a too slow response time but, by contrast, it never fails. Other experiences are performed in order to use the infinite chain of atoms which constitutes the DNA molecule as a storage media and the prospects are quite fabulous!

Obviously, such "machines" would be, by nature, biocompatible. Only one more step is required to emphasize a monster, an "equivalent brain" biologically created or associated to a "normal" one to acquire all the functions which are inaccessible to the pure logic and Frankenstein is already knocking at our door!

The intelligent computer

How much progress has been made towards responding this old question: "The computer is a stupid machine which can

---

[23]Luke Muehlhauser and Anna Salamon, *The Singularity Hypothesis: A Scientific and Philosophical Assessment*, Springer Ed., Berlin (à paraitre).

never compete with human brain?" No one would disagree however that the computer is clearly superior in logical, mechanical operations as calculation for instance. But our "smart mind", we said with complacency, has nothing in common with a machine!

One remembers "Big Blue" the IBM computer which successfully fought with the chess champion Kasparov in 1996. Since this challenge, "DeepFritz" software installed on a conventional PC beat a defending champion Vladimir Kramnik. One still said "yes, but chess is mechanical, simple logic, creative intuition is proper to the man". Yes of course but ...

In May 2011, IBM has gone again with "Jeopardy" a very well-known game in the US, similar to the French *"Question pour un champion"*. The challenge was that the computer had to behave exactly in the same way a man would do and without cheating: it had to understand the question, vocally formulated, look for and evaluate the possible answers, sort the best one, and answer with a synthesized voice; all of this had to be performed faster than the two selected challengers.

The computer was represented by a desk between the two men. The organizing team waived the idea of giving it a human look, as Japanese certainly would have done. Psychologists indeed agreed that, for us in the West, there is a spontaneous repulsion regarding a humanoid robot whereas a simple raw machine is more easily accepted would it be intelligent. Would this mean that a kind of "racism" is to be expected to rise against robots?

The final question was: "Who is Bram Stoker?" and the answer was "Bram Stoker was an Irish novelist who, in 1897, invented the legendary and bloodthirsty character Count Dracula and made it a best seller". All of that did not prevent the computer (dubbed Watson) to win a million dollar against the competitors who were selected as the best in the game.

This performance lies very close to the Turing test which is to evaluate the competition between man and machine.[24] The computer, now, really enters our life as a real partner; today, a "simple" phone is able to understand and answer the questions as a human would do. All of that in a single "chip"!

Of course, one would say again that it is only a matter of "correspondence" there is no consciousness lying behind. Certainly, but one starts talking of "formal logic", that is to say software which manipulate not only data but also abstract "concepts". The more striking progresses involved mathematics: algebra, number theory, groups etc. ... Software has already been proposed for small computers which give not only a numerical answer or graph but readily the analytical solution to the problems of the classical mathematics. Students then no longer have to write pages of calculations to reach the solution of the differential equation or the polynomial system; they will get the solution in a click!

---

[24]The first fulfillment of the Turing's test just happened on June 7, 2014, anniversary date of his death. The winner is a Russian computer ("chat-box") which took the personality of a young Ukrainian. At the end of a 5 minutes phone call, 30% of the English judges strongly felt they were talking with a real human.

But, incidentally, with the problem of "singularity", following Kurzweil, it would appear that people would accept more easily the intelligent computer (providing it would not externally resemble a human!) than the possibility of an extended life.

Simulation and virtual worlds

At the boundaries of the AI and robotics, there is a domain about what less is said although it is of an essential importance: the specific domain of simulation. Applications are to be found everywhere and at various levels of sophistication but they represent an open door to the virtual worlds which are to play a growing role in our societies. Suffice it to say that the increasing appetite the teens develop for this kind of applications which exceed the level of a simple "role game".

Simulation starts with the realistic replication of the real world in the aim of training, education or reflex drill in a given context: driving a car, piloting a fighter plane or a 500,000 ton tanker. Everything is made to recreate a visual, acoustical or mechanical environment of the situation. The realism generated by the computer will be pushed to such a limit that the illusion is fulfilled. But very rapidly the research has overshot the educative purpose to diverge towards pure "artistic" creation, dream or even ... psychiatric therapy.

The Japanese Riken research center even developed a special "helmet" for a total immersion in a world where reality and virtual creation are mixed[25] to lead to an interface man/computer. To the point it can get fully confusing!

This also helps to appreciate physical, chemical or biological inaccessible contexts up to the limit without taking any risk. The credibility of the result only depends on the faithfulness of the simulation to replicate the process. So it is easier to simulate a nuclear bomb explosion than make it for real; so it is more reassuring to artificially simulate a chirurgical operation of the brain than putting it into a real practice on a guinea pig.

Japanese (even them) have just realized on a super computer, a full simulation of the heart the precision of which is claimed at the molecular level. This also is accompanied with the corresponding imaging facilities. Every critical situation can be emulated to appreciate the evolution and address the issues.

Simulation is now widely accepted in the industries, it is integrated as a normal prerequisite and no longer as *a posteriori* verification in the elaboration of a prototype. This to the point that there is an inversion to be observed in the behavior: people are more confident in the virtual prototype which becomes "more real" than the real one! One is more confident in the simulator!

A story can be reported which is quite illustrative of it: when the giant Airbus A380 was to be created, they first made a simulator and the future pilot got trained on it before the real plane was completed. The result was that the pilot,

---

[25]This to be called substitutional reality, enough to say!

upon return from the first flight, gave this naïve and sincere comment: "I did not have any surprise; it behaves exactly as the simulator." This was a splendid vote of confidence.

In the same order, the new fighter F35 of the US Air Force will be a single-seater craft; a two-seater version for training is not pending, the training will be exclusively performed in a simulator and the pilot will be directly operational on the combat version.

The day an "artificial" man would be created (some are already thinking about it) it will be mandatory to first build a software reliable model, the exact realization will come later as a matter of course. In the present growing complexity of our world, the human performance that will be necessary in the company may not cope with the demand. Michel Serre advises us: "new technologies are bound to intelligence!"; a whole program!

The man alone will likely be unable to face the problem, he will need a support. A support all the more necessary than the millenary solution to invoke god(s) will be erased. Will the machine be the solution? Or may we be led to help man to mute biologically?

...

Chapter 4

The Biological World

Beyond all the advances in physics and technology, evolutions must be taken into account that have been introduced by these progresses in the understanding of the living world, beginning with the smallest fundamental life units: the cells.

In the exuberant jungle of the microscopic world, the biologist had to invent a specific and dizzying vocabulary, in the wake of the diversity of the problems. Only insiders will be able to find their way in this environment where chemistry gets sublimated.

The "original soup" coming out from the early ages, peppered by amino acids may be originating from the deep space[1], spent billions of years creating, by chance, more or less complex, more or less strong, more or less sustainable molecules. These molecules spontaneously (?) assembled and disassembled and finally gave rise to the first viable structure,

[1]Maybe Rosetta and Philae are now to give an answer to this question?

the first "living being": the cell named LUCA[2] and its first hereditary message: the DNA.[3]

Billion years continuėd to scroll, generating the fantastic diversity of cells we are able to see nowadays: size, form, function, metabolism or aptitude diversity. These unicellular organisms, bacteria, bacillus etc. ... all benefit the same basic property of life: the ability to reproduce and its corollary: the death. When we say "all", it obviously suggests (scarce) exceptions: stem cells among others do not die naturally.

After the mono-cellular stage, the building of more complex structures started giving, step by step and following a natural selection and specialization, more elaborated assemblies always governed by their DNA support and the same laws of adaptation to the environment, thus giving birth to the animal and vegetal present world.

The story of the microscope

When A. van Leuwenhoek (1676) built his first microscope, he looked at a drop of murky water taken out of the nearby pond. You can imagine his surprise to see plenty of microscopic living beings, some 10 micron large[4] sculling together.

Nobody, at that time, would have imagined the possibility of such a world of living things that even the Sacred Scriptures

---

[2]LUCA: *Last Universal Common Ancestor.*

[3]DNA: Deoxyribonucleic acid.

[4]1 micron = 1 thousandth of a millimeter or thousand nanometers.

did not mention. Very disturbing breach! The Man was just discovering a universe that God had hidden to him.

This discovery was as difficult to be accepted as will be the negation of the spontaneous generation which the same Leuwenhoek also discovered later with the corresponding spermatozoids. Some years later Robert Hook, with the same microscope, showed to the world the existence of even more smaller elements: the living cells. A mysterious, unexpected and worrying world as the nanoworld we are presently discovering was just to arise.

These "animalcules" (or "rotifers") Leuwenhoek discovered are the smallest existing "animals" at the limits of molecular chemistry. They are made of 1000 cells and 16 neurons, they are able to swim, eat, and move freely. It appeared that they had a clear life of their own which is managed by a complex biochemistry; they lived and died in colonies of individuals. The numerous species are gendered or not (the "bdelloids"). In the last case they are especially resistant and can survive the worst possible treatments.

Leuwenhoek then discovered that they could be fully dried and indefinitely stored; a simple drop of water will be enough to bring them back to life. To reproduce, they clone (replicate) by themselves through parthenogenesis. Their genetic material, then, is maintained and they are unable to evolve, but they are immortal. This example clearly demonstrates the correlation between sex and immortality.

But things got more tricky when it was discovered that such tiny and individual cells were also able to stick together in order to give rise to larger entities, each cell transforming to play a specific role helpful in a common welfare, in an organized collectivity: the tissues of the living which gave rise, in a definitive step towards, to the human being.

This discovery of our cellular nature, as every scientific news, has been shocking in the first glance but it soon appeared that such a coordination, such an underlying organization should require a fabulous "conductor" and could not arise from a simple game of chance but unquestionably from a superior divine authority. Religion was, then, able to swallow the new discoveries because all creatures, as small as they could be, are God's creations. Same thing occurred (but with much more reluctance) when the theory of evolution came to participate in the landscape of our origins.

Similar to the present scenario, everything was done in opposition to the established opinion, to the religious belief, and with the use of a new instrument nobody clearly knows about the "whys" and the "hows". It was just the premises of the modern era. Disbelief and fear of the unknown were exactly the same, as well as fierce oppositions and timely excommunications; one frequently used to refer to a Science where the Devil was to play his role. Why would there be the need to search for such disturbing truths?

It can be still bearded in mind through this adventure that the optical microscope will remain for the coming centuries the essential and indispensable instrument for the exploration of the microscopic world. Of course, many improvements were brought in the basic optical operation of the system (fluorescence microscopes) before some new kinds of instruments came to help and improve the performances such as the beam Scanning Electron Microscope (SEM) or now the Atomic Force Microscope (AFM).

But many other means have recently appeared to help the biologist get still more information: centrifugation, liquid phase chromatography, electrophoresis, radioactive markers, and flux cytometry, also with the new tools of the genetics. Given such a profusion of tools, nobody would wonder that the improvement in our knowledge in the domain of biology has really rocketed during the recent years.

Life and death of the cells

Cells, notwithstanding their complexity, still are even the simplest of the organisms. The understanding of the entire living world firstly begins with a detailed evaluation of the biochemical mechanisms which govern the functioning of the individual cells. We, now, begin to understand many details but numerous questions are still pending.

Researches have flourished in every direction from the simple observation of the behaviors towards more and more refined experiences as more and more sophisticated instruments (the key element) and working methods were implemented; "autonomous" cellular organisms were investigated as well as those which are integrated in living animal or vegetal organisms.

➤ *Autonomous cells: bacteria and coli–bacillus*

These mono-cellular organisms are of a wide diversity of size (from 1 to 10 microns long[5]), shape, metabolism, and ability to move. Bacteria can colonize under every environment even the most hostile; some anaerobic bacteria can even do without oxygen. They originally accepted, without any problem, the original "pre-biotic" soup which was sufficient for their proliferation.

The most studied bacteria is a prokaryotic coli–bacillus, the well-known *Escherichia coli*. Its reproduction cycle is very short (3 generations in an hour) which makes its study easier.

These primitive beings benefit astonishing properties which we are now trying to decipher because from them, one might be able to deduce many immediate therapeutic applications useful in the current healthcare sector.

Some cells (a dozen of species) such as the well-known very common bacteria *Deinococcus radiodurans* are recognized to be immortal; they are able to come back to life after years of lethargy without any food or water, in the hot sands of a sun burned desert. These bacteria were also dubbed as "corned beef bacteria"[6]; they are capable to fully recover after being put into pieces by massive doses of irradiation. Its fragmented DNA spontaneously reassembles in the right order.

---

[5]We are, in any case, in the micro-world, not the nano-one.

[6]After WWII, corned beef boxes of the military were sterilized with a strong dose of gamma rays and it was observed that this bacteria (on the other hand harmless) withstanded all these treatments.

This strange behavior is now understood: it comes from the fact that the cell contains not a single strand of DNA, but several. When receiving the radiations these strands are cut randomly which makes the reassembling easier and in the right order because they have been cut at different places.

In any way the proliferation of these mono-cellular organisms only depends on the presence of the available nutrient. It has just been discovered that their death is directly related to the oxidation of the proteins which take care of the DNA.[7]

It can also be added that bacteria generally do not have gender and this seems to play a key role in the longevity. This property is also involved in their resistance to desiccation and radiation effects, the basic explanation being that the acquired genes have to be preserved from a sexual mixing.

➢ *Specialized cells: the worm, the fly and the little mouse*

The animal world is especially diversified and variable; the study of all the species at the same time would have been a too much large and embarrassing enterprise before reaching conclusions which could be extrapolated to our essential center of interest: the Man. The biologist then selected species each typical of a size category and ease of experimentation. Beginning with the smallest, they chose a minuscule worm, then a fly and finally a mouse in an increasing complexity order. These species have the advantage of being of an easy

---

[7] M Radman and A Krisko, *Journal of Science Academy*, 2011.

access and to reproduce very fast thus allowing a quick evaluation of the influence of the treatment on the intergenerational consequences.

- The worm *nematode caenorhabditis elegans*

This is a small worm which has a mean life of 15 days and which generously reproduces. In a hostile environment, it transforms into a mutant which resists the worst conditions. On this worm, genes have been identified which influence the mean life of the species.[8] An American lab even succeeded by genetic mutation to extend its mean life by five times. If a man extrapolation was possible, then it would make us able to live for 500 years! Quite a lot!

- The vinegar fly: *Drosophila melanogaster*

The more evolved condition of the fly allows the study of the successive stages of the shaping and specialization of its organs. This fly was at the origin of the chromosome theories of the heredity (Mendel's laws). Flies were also the first transgenic animals to have been produced because it has allowed to establish the first map of a genome which describes the correspondence between the genes and the organs. Today the study of its "innate" immunity system leads Jules Hoffman to the Nobel Prize.

---

[8]Buck Institute for Research on Aging: http://dailygeekshow.com/2014/03/06/des-scientifiques-ont-decouvert-que-letre-humain-pourrait-vivre-jusqua-500-ans/.

○ *The mouse Mus musculus*

The mouse gets closer to the man with its two sexual chromosomes. In an embryonary stage the formed initial cellular mass contains stem cells. A foreign DNA can be injected which will be assimilated, thus making it possible to get an approach of the mechanisms underlying the functioning of the nervous or the immunity system. This is very important in the study of the human diseases.

With these three investigation supports it becomes possible to observe the fundamental mechanisms which drive the life (and the death) of these cells the living is made of. Unlike the cell organisms, the proliferation mechanisms, there, do not exclusively depend on the nutrient but also on a severe regulation by proper mechanisms.

The morphogenesis of a complex organ results from a differentiation of the tissues. From an organ to the other there do are common points but also notorious differences; this study still is in a very beginning. One should like to understand how a lizard restores its damaged tail. It is the only case in the mammal category where cells are capable of a spontaneous rejuvenation of a complete organ. What is the secret, how do these cells resist the oxidation? Researchers are working on ways to regenerate trachea or even hearts by such a manner. Skin, as blood, is in a constant renewal but the operation is limited to a single tissue, not a diversified organ.

One knows that prokaryotic cells communicate; external molecular signals regulate their behavior and sometimes trigger the death, as a function of the needs and the equilibrium of the whole organ.

Cooperation and a specialization do exist in the cell's division which results in the elaboration of a particular tissue.[9] The first ranking cells at the origin of an organ benefit of exclusive properties: they are the famous stem cells.

An important research domain still poorly cleared is devoted to the death of the cells. It takes two different directions. There is a programmed death[10] (or apoptosis) of these special cells and the "disordered" death (or necrosis). In the first case, death occurs deliberately to make it possible for the tissue to renew or redesign; it is necessary and one could say "agreed". In the second case, the death is "accidental", unexpected in the program; the cell is then torn into pieces which are "digested" by the neighboring cells.

Another phenomenon can also be added which is not clearly understood up to now, but this is very important point: it is named "cell senescence". From a certain stage, some cells stop proliferating without knowing exactly why. The total number of possible divisions is intrinsically limited.[11] Tumor cells, for their own part, are immortal and reproduce indefinitely, that will at the end induce the death of the organ (and the whole body).

Nevertheless, huge leaps forward have multiplied: Jean-Marc Lemaitre succeeded in restoring its youth to a

---

[9] 200 different types of tissues are identified in the mammal's bodies.

[10] It is also called "cellular suicide".

[11] In the case of the skin cells this number is between 70 and 80. It is called the Hayflick limit.

century-old human cell with the introduction of new genes; no malfunctioning was to be observed with respect to the young ones on the first day of their life. That opens wide and fabulous horizons for the regenerative medicine. He even added that, actually: "the wear of time has been removed ... one can now imagine being able to erase the diseases related to the age!" It is true that the genetic engineering gives a decisive contribution and we still are in the very beginning.

This is quite in line with the concerns of the English biologist, Aubrey de Grey[12] who considers aging as a built up of cellular damages naturally resulting from a normal biological activity. These damages would be curable if we pay special attention to them. Each time a cell divides, its "telomeres" shorten until a complete exhaustion is reached. But it does contain an enzyme, the telomerase, which is able to reverse the process and that is what lengthens the cancer cell's life.

Researchers at Harvard succeeded in making normal cells take advantage of such a treatment.[13] The key is in the proteins; a better knowledge of their properties will result in a better control of their degradation, that is to say their oxidation. One knows, until 2010, that the degradation of the DNA itself is not involved in the process.

A curious element also dwelled the attention of the biologists, it is the case of the naked "mole-rat"[14] (I am awaiting

[12] Foundation: *Strategies for engineered negligible senescence*. Cambridge, UK.

[13] *Nature*, November 2011.

[14] In short, RTN for French people.

for a right translation!). This strange animal is also named *Heterocephalus glaber*; it is especially ugly but it can live over 30 years, that is three times a normal rat life and, on top of that, it does not show any senescence phenomenon. This exceptional endurance is a corollary of its total immunity to cancer.[15] It seems that the very origin of that could be found in two protein molecules[16] which associate in order to inhibit the contact between cells and thus prevent an anarchic development. One more idea to be explored!

However, if the aging is to be stopped or slowed for humans, a single magic recipe would not be enough because the enemy is numerous and diverse and it would certainly fight using unpredictable secondary effects. As Laurent Alexandre says: "We should need a cocktail of therapies, associating several weapons coming from GNR … one could not become immortal tinkering around with a single gene!"

Stem cells

Stem cells are obviously a challenging area of research, a domain which is special and booming since a decade, in spite of banning or ethical oppositions, in the occidental countries at least.

At the start are the embryonary stem cells: they are called "pluripotent" which means that they are at the origin of the creation of any type of specialized cells in the human body. This property is particularly interesting because anything can

---

[15] Science et Avenir, 746, 2009.

[16] *Proceedings of the National Academy of Sciences* (27 October 2012).

be obtained there from. But, in this order, it is required to "guide" them accurately during the evolution, otherwise they turn into wild unusable proliferation or a tumor.

Stem cells let abandoned, with nutrients in a Petri vessel, will give rise to a shapeless mass combining, disorderly, hair, teeth, and other unfinished tissues. But when controlled they selectively can bring to lever, heart, and skin (wrinkle-free) tissues.

From these immortal cells results a cascade of daughter cells more and more specialized, until the final stem cell of the selected tissue from which the following dedicated daughters are mortals. A cell to become a cancer cell has to de-specialize and then it becomes immortal.

These stem cells are the basic cells which are the subject for the most drastic experimental, religious or political prohibitions. Indeed they originate in four days embryos and some say "we are destroying a life in the hope of saving another one". Such a moral standard stays in a country (France) where abortion is legal and even reimbursed by the social security, but that's it! People in the Asian countries do not care so much and the works, there, are moving ahead quickly, without ethical constraints. GW Bush closed the financing, B Obama reopened them, M Romney during his election campaign said yes and then no! The debate is on!

However, any regulation can durably resist to such an attractive transgression: news was just released that the French law[17] had become a little more permissive in respect

[17] Senate Decision, November 2012.

of stem cells, the study of which is now allowed (under many restrictions, but the door stays open).

For some time now, Robert Lanza,[18] a pioneer of these techniques, knows how to reproduce embryos *in vitro,* then, from a single fertilization embryo, it made it possible to obtain a countless generation, keeping alive the initial one. Nevertheless, it has also become possible, in some cases, to use specialized stem cells from the saliva or adipose tissues and thus get free of the embryonic stem cells.

More recently[19] one has even succeeded in inversing the evolution cycle and get back to the standard embryonic stem cells from specialized skin cells. Everything would be fine since embryonic stem cells are no longer required for the future investigations. Except that the American Catholic Church still opposes to these studies arguing that it is an open door to human cloning; which is not so false!

Professors John Gurdon and Yamanaka Shinya were award winners in the Nobel Prize for their contribution to the genetic reprogramming of cells to get rid of the original cell. We are still a long way from a magic therapy against every ill in this new medical paradigm but:

— Two women in Los Angeles (UCLA[20]) who suffered from AMD had restored their sight after an eye injection of 50,000 embryonary stem cells. It is thought that within

---

[18] *Chief Scientist, Advanced Cell Technology* (ACT), California.

[19] Mai 15 2013, Portland University (Or).

[20] See *The Lancet*; or *Fortune*, October 8, 2012.

five years, such a therapy could be as common as corneal transplants. Testing is now underway in the US and in England on tens of patients.

— Mice have also been successfully treated against cancer, Parkinson, diabetes, stroke etc. ... Extension to the man's case is to be considered and could be a phenomenal step forward.

In the US, the major obstacle up to now has been the difficulty of funding these researches because of the resistance of both the politics and the public[21] as well. Wall Street was reluctant. Such researches are long and costly for a return on profit still unclear up to now. People talk about a "monetary purgatory" between the lab and the moment the validated product is placed on the market.

However, things have changed recently, funding are to be released and important donations also came. Los Angeles and San Francisco Universities have already received important funding; they created dedicated laboratories, and they hired the best experts. Clinical tests on a large scale are to be launched (diabetes type1, leukemia, stroke, heart etc. ...). Scientists says they have reached the point of roll-over of the tendency and that a powerful start-up is on the way, induced by the abundance of new and performing instruments (always them) now available: genetic sequencing, imaging facilities, automatic cell sorting etc. ... Within 10 years all

---

[21] Similar resistances were also found in France.

these therapies will be operated in a routine duty predicts Arnold Krigstein director of the UCSF Center … if, however, these novelties do not come sooner from China.

Our morals violently opposed against the human cloning; that had in no way prevented researches to proceed among the Asians who do not obey the same taboos. However, the "partial" cloning could be perfectly accepted and it is assuredly tested.

Let us imagine somebody whose heart (or another vital organ) is at a conclusion stage of aging. It is quite conceivable to grow muscular heart stem cells *in vitro* and inject them in the heart in order for them to proliferate and, one step at a time, take the place of the older; they will clone, that way, an all-new heart.[22]

This has been experienced already at Saint Louis Hospital (Paris) by J Laghero on July 2013. Embryonary stem cells have been "amplified" *in vitro* and then engaged in the specific track of the heart cells. A "bandage" was made with these cells directly on the heart in order to regenerate it.

This way of doing (if it really works) will clearly be better than the heart transplant because, in this case, there will not be any problem with the immunity system of the patient, the DNA being the same. What kind of morals could disagree?

A similar test has also been successfully performed with a retina which nevertheless is an especially more complex organ.

---

[22] Le Point, July 30, 2013.

To go even further in the same domain of the cells, the American food industry very seriously emphasizes to grow artificial meat, cloned *in vitro* (for instance artificial sirloin steaks!). That would be very cost effective and would avoid growing cows; the process is faster, cheaper, without waste and cuts ... and also keeping rid of the methane (then CO_2) which the cows are largely providing us!

The latest news tells us that such a "meat" for hamburgers is already available.[23] It has been officially tested in a "grand restaurant" in London (June 2013) by a distinguished assembly of gourmets[24] (no French "gourmet" among them!) who testified it was tasteful! The only thing that remains is to finalize an adapted economic model before launching an industrial production.

Finally, to conclude with the cells, a French lab took over the Swedish technique which allows reconstructing a cornea and the epithelium of a blind person with the help of stem cells taken in the mouth and grown over a special Japanese substrate. Then corneal implant is no longer a topical issue!

Preventing the death of the cells

Apart from the cell implants used to reconstitute a failing organ, it could be imagined to directly reconstruct the affected proteins, but there is also another strategy which would

---

[23] B Walsh, *Grow a burger*, *Time*, March 25, 2013.

[24] www.podcsatscience.fm/dossiers/2013/06/15/les-steaks-de-cellules-souches/.

consist in preventing the failure, that is to say the aging or the corrosion of the cell. We already understood[25] what has to be done: prevent the free-radicals to become aggressive and evacuate the damaged materials. This can be obtained by controlling the genes which are responsible of the "silent mutations". There were some serious avenues to be explored by the bio-gerontologists.

Generally speaking, the doctors are rather reluctant to the changes and to the advances of the Science which trouble their certainties. This attitude is in some way a good thing for the patients: it protects against dangerous over-enthusiasm in the care they provide us with. But it is also an impediment for the evolution of the medical treatments towards more efficient therapies.

For instance, neurologists often develop a furious pessimism in the future of any therapy against Alzheimer disease. Similarly, gerontologists show a similar defeatism and refuse the new ideas in fighting against aging.

Then, it must be recalled that animal species do exist which can be termed immortals (hydra or some jellyfish) and we have no way of knowing very well why. Their studies have just begun. Let us wait a bit more!

Genetics

Some words in order to come up with biologist language that would be unacceptable to all. Everything revolves around this

[25] Miroslav Radman, Au-delà de nos limites biologiques, Plon, 2011.

huge double helical molecule which James Watson discovered some 60 years ago and which carries all our hereditary material: the genome. This molecule is a polymer of nucleotides, infinitely diversified from four elementary bricks (or basis), molecules named A, T, C, G. They are assembled in pairs to form groups of four representing each step of the spiral structure. The total length of a human DNA is in the range of 8 cm, I let you make the calculation of the million basis involved. It is somewhat like the "Great Wall"[26]: it is very long but at the same time so thin!

When a cell is to divide, first there is a phase of replication of the DNA molecule as a destination of the daughter cell to grow. However, it happens that the copy is not carefully copied, this is the issue of the cellular mutation. These genetic mutations also happen naturally during sexual reproduction where the DNA strands are shared to give rise to one single, then providing biodiversity. Such molecular mutations may also be artificially induced, giving rise to a new type of organism which we now call a GMO.[27]

This DNA, also with associated proteins, constitutes the chromosomes shaped as an X or a batonnet in the nucleus of the cell. We all have our own DNA originating from our father and our mother ... plus a pinch of chance. It is this special arrangement which makes our genetic message. We all possess 23 pairs of chromosomes one of which is sexual.

---

[26]This is the reason why the Great Wall cannot be seen from the moon contrary to a widespread opinion.

[27]Genetically Modified Organism.

Gene sequencing

Obviously the analysis, even partial, of the constitution of a DNA molecule is a monumental work; to the point that Watson, the discoverer, did say during the first attempt (around 1990) that such a decoding would take a thousand years!

Within five years, the job was done by Craig Venter for a bacteria and in 2003 the first human sequencing was achieved … onto a sample from Venter himself! Total cost of the venture, 2.7 billion dollars; duration, a dozen years.

Today, 11 years after this first success the same is done for only 1000 dollars and IBM[28] announces it will be able to do it soon for 100 dollars and in 10 minutes! One even succeeded in getting the DNA of a Neandertalian! The cost of sequencing is reduced by 50% every six months!

However, all the genes are not yet deciphered or related to specific identified effects, many of them do not seem to play any direct role[29], but knowledge advances at an incredible pace. We may restrict ourselves to a research on the only 200 genes which are clearly related to recognized risks, the cost for such an analysis drops to $99. Of course, the analysis can be ordered through Internet!

---

[28]One could be surprised to find IBM in this biological venture, but on one hand the sequencing itself specifically favors software and computation and on the other hand the following use of this tremendous amount of information obtained from millions of persons will require an enormous data crunching which the computer is the only way to achieve.

[29]This point of view was recently questioned. It seems now that they do play a complex and indirect role.

When the first sequencing was performed, Bill Clinton, then President of the US decided that the process will be made public and that no patent should be filed because it is a matter of the Humanity heritage. But it still remains that the implementation of sequencing has remained regulated in some countries (like France).

At the very beginning of the story, France was in the front of the race for sequencing, with "Généthon" and the team of Daniel Cohen and Jean Weissenbach who produced the first partial map of the genome in 1992. But the funding (too much limited for such a project) was not comparable to that of an "American dream". Nevertheless, Généthon has received the Galien's award for its contribution in the genic therapy. From that time on, the progresses essentially followed the evolution of the computers and, currently, tremendous efforts are made in the US.[30]

How is the genome analysis performed in determining the arrangement of millions of basis, if a complete operation is required?

Initially, two different methods have been proposed:

— One uses the synthesis of a strand identical to the model (Sanger's method).
— Another one, on the contrary, uses the destruction of the model, piece by piece (Gilbert's method).

---

[30] Jean Weissenbach: private communication.

Only the computer is able to replace such a bunch of data in the right order.

Other more radical methods have also been proposed: such as the "Shotgun" method of C Venter, the NGS (Next Generation Sequencing), and this clearly shows that there is a constant evolution towards faster programs, cheaper, more performing and more secure procedures. The emerging techniques of fluorescent markers were a key step in this onward rush.

These programmable DNA sequencers are costly and complex systems but they are also more and more readily accessible and performing. The technology tirelessly progresses; it is now feasible to analyze a million bases in a second. Today, everything can be known of an individual from a mere sample of saliva!

Sequences are becoming increasingly better identified and confidently related to the corresponding effects. It becomes possible to begin to target particular genes and thus modify their impact; it is also possible to add new genes the presence of which is supposed to be beneficial.

To allow comparative studies, a standard reference for a "mean" genome has been established which is called "HuRref19"[31] and which is to replace the previous one, the Venter's one.

However, we must not derive from the genome analysis, a definitive deterministic character. In the evolution of the

---

[31] For: Human Reference 19.

cells, the tissues and the complete bodies, other parameters are to play a decisive and significant role, such as environmental or education factors.

The genes suggest a predisposition, a weakness, or a "knack", which can as well never occur if the external factors do not contribute. Somebody sensitive to the melanoma cancer might be less likely affected if he spends his life in Alaska rather than in Sahara! There will always remain an uncrackable part of uncertainty in such analysis ... but this part is shrinking fast as researches are progressing.

The most important domain which is reportedly concerned is of course cancer and we would like to reach a natural and transmissible immunity from a targeted genic therapy. But, considering the huge diversity of individual situations, such a therapy would not be unique and universal, on the contrary it should be multiple and customized. This is a price to pay for a predictive and preventive medicine. The initial testing in this way took place in 2000 on two "boys in the bubble" by A Fischer with the help of retroviral carriers. From that time on, the progresses are obvious but, still then, exploratory.

Some countries (like France) are afraid of the generalized use of such analysis and the following public disclosure of the genetic intimacy of the persons. This could lead to serious social drifts in many domains such as hiring process, segregation, access to insurance and others. Then, DNA analysis is definitely limited to much more specified situations such as medical specific cases or police investigations. However, everything is reachable on Internet at a moderate cost! In our times the precautionary principle cannot hold for a long time. It can be expected that, in a near future, DNA analysis will be as common place as a simple blood test.

From year 2000, genetic manipulations never stopped with explorations, manipulations and results scattered in all directions. We shall see some of them, below, among a whole host of fantastic discoveries.

Some says that Craig Venter, as God, has "created life". In fact on May 20, 2010 he succeeded in reintroducing a DNA molecule in bacterium the original DNA of which has been removed. The new DNA was designed in a computer and carried an artificial genome conceived on purpose. The new bacteria[32] had a normal life and reproduced in the same way as any bacteria with its new genome. Joel de Rosnay[33] (Head of the Forecasting Service at "Institut des Sciences de la Villette", Paris) said: "It is the first living being which does not have any ancestor!"; this is because the "father" was a computer!

In fact, we should better say "Craig Venter has redeveloped life" as the modification has been introduced in a living bacteria which did exist before. However, tentative research is currently going on to create living cells from scratch, from mineral. We already know how to make the membrane of the cell; the rest is just about to arrive, soon!

To be also noted, the tireless Craig Venter just launched an industrial 800 million dollar venture to create artificial

---

[32]Referenced "JCVI-syn1.0".

[33]Joel de Rosnay, 2020 les scénarios du futur, Fayard, 2007.

optimized bacteria and algae for producing a cheap bio-carburant.

All along this adventure in genomics, we may hope to find soon some genes which could extend life and provide a protection against aging and diseases. Nevertheless, it has never been possible to find a specific gene responsible of the death. Death is fundamentally not programmed by Nature. It comes from a normal wear[34] against which the Nature renounces keeping up its protection after the procreation delay. This "wear" depends on the individual, following his genetic predispositions, his own life and the environment.

The genetic sequencing is then a major advancement in the knowledge and manipulation of our physical "self". An important progress is to be assuredly expected in the way of achieving a good health and to age well. Anyway, every progress always comes with new problems to be solved before getting in the public domain of every day.

This genetic analysis, when fully completed, will say everything about the risks incurred which we will have to prevent; but is it always a good thing to be informed of possible threats? Would we always be able to supply the adapted medical response? Will we have to live with the stress of a disease which could possibly never develop? Will the doctor be obliged to say everything to his patient (this is an old mantra)? Will the patient be able to learn everything by himself (Internet)? Prenatal DNA analysis of the baby is to become

[34] This wear is coming from the oxidation of the enzyme molecules.

common if not mandatory; what if the results obtained simply dissatisfy the parents? Will we have to communicate the information or restrain it?

GMO

The almighty power of the mastery of the possibility to change the genetic composition of living being upon request started with the vegetal domain. This led to the generation of a violent social problem: the GMOs. The oppositions were strong and diverse, the arguments often scientifically irrelevant and impassioned (when not dealing with deliberately forged documents), particularly in Europe where the persistent precautionary principle prevails. However, the economic stakes and the necessities cannot lead, deliberately, to reject the GMO *a priori*, without a careful prior and serious examination of the problem case by case. Transgenic corn is commonly used in the US for 25 years without any trouble. These can be found in meat-and-bonemeal used in Europe for animals!

Populations to be fed grow without limits and the culture environment has to be controlled for a better optimization. The genetic modifications are the underlying foundation of every living being and they are purely random; natural selection allows putting things right if the result is not satisfying. Our famous José Bové himself is a random GMO, as everybody, and he cannot do anything about it.

Other nations are involved in this field and nothing would stop the advancement of the research because the taboos and the red lines are not everywhere the same. Chinese today outperformed the US in the number of publications in the domain of gene-technology.

The accumulation of genetic data from everybody on digital servers will give rise to a huge data base, particularly if the medical history of everyone is stored in order to allow cross correlations and individual monitoring. This enormous archiving machinery will also make it possible to enter in a critical evaluation of the drug efficiencies and their cross-interactions when several drugs are used at the same time. There will be a colossal work of data crunching to be performed.

Cells and cancer

Among the many immediate goals of the Science there is one which occupies a priority place, it is Cancer.[35] At the origin of so many deaths it constitutes a leading obstacle to life extension because aged persons are mainly concerned.

Its origin is to be found in a cellular deregulation which is induced through a genetic perturbation. An oncogene cellular clone is formed from an alteration of 10 to 20 specific genes which blocks the normal death (apoptosis) and then makes the affected cell immortal. An uncontrolled proliferation follows which deteriorates the surrounding tissue and propagates.

Such a proliferation is to be reinforced by the migration of the cancer cells towards other organs (metastasis), through

---

[35] In France, it is the first cause of mortality.

the blood stream and lymph for instance. Evolution patterns and processes are largely diversified which makes adapting a targeted therapy as more difficult to achieve; this, all the more, as the concerned organs are also of a wide diversity: lungs, breast, colon, pancreas, prostate etc. ... with, in each case, cells whose properties are different.

Obviously, this diversity substantially increases the difficulty in the elaboration of adapted therapies. Efforts were multiplied to select typical situations and adapt targeted replies. Recent advances in the genetic analysis constitute an important step forward in the understanding of the source of the plague.

Up to now each discipline was working for its own behalf and with a limited coordination with the others, which did not make things easier. In spite of that the proposed treatments (surgery, chemo-therapy, radiotherapy, even genetics...) have become more and more precise and have led to an evident reduction of the deaths; however, we still are far from an achieved eradication.

Currently, a new very aggressive approach has just been implemented in the US which, for the first time, put in place a direct collaboration of teams from different specialties in a close cooperation program.

An organization called SU2C,[36] directed by the Nobel Prize genetician Phillip Sharp, gathered considerable resources and mobilized the most talented specialists in the

---

[36] For: *Stand Up To Cancer.*

different fields so as to work together. It makes it a true "commando"[37] operation where sharing actions and results succeeded to apply in spite of divergent industrial interests, confidentiality exclusiveness or patent rivalries; every barrier has been removed.

A considerable pressure is put on the researchers; the transition from the lab to the clinical test, until then so lengthy, is reduced to nothing, the purpose being to reach, in a very short time, the good agreement between the detected gene mutations and the proper combination of drugs to be used in each specific case. Collected cancer cells are simultaneously analyzed in five different laboratories in order to identify the process with precision and timeliness.

The program was dubbed "Moon Shot" in an analogy with the lunar program initiated by JF Kennedy. The teams are evaluated from the results on the patient and the distributed financing consequently follows. For the first time, genetician, pathologists, biochemists, bio-statisticians, surgeons, oncologists, even software specialists are joined together also with nurses and technicians who learn to understand and to speak to each other. This makes it a striking example of the effective convergence of the sciences. Hundreds of drugs and markers are under elaboration and evaluation in this operation.

A chip has also been developed which is able to selectively "trap" cancer cells which happen to circulate in the blood ves-

---

[37] Bill Saporito, *The conspiracy to end cancer*, *Time*, April 2013.

sels and could induce metastases; it thus brings an early warning system for a preventive and immediate treatment.

No doubt such efforts will rapidly bear fruits and lead to significant advances in the fight against cancer.[38] A success in this domain would certainly allow a substantial step forward toward our longevity.

About mice and men

In 2009, an Asian laboratory succeeded in synthesizing "artificial" spermatozoids the DNA of which were computer programmed. The corresponding mice were absolutely normal! Of course, in this example the corresponding eggs were natural but it has been promised to make artificial ones very soon because we now know how to collect ovarian stem cells.

Still better than that: one has succeeded in generating bimaternal mice (from two mothers at a time) and, surprisingly these mice have the characteristic of benefiting an extended life clearly longer than the others! We find here too a definite relationship with sex and longevity.[39] This will give hope to "mono-sexual" couples who are waiting for that! We have already directly transgressed the essential laws of the nature.

This mice experiment has just been extended to humans. During the 2013[40] summer, 30 babies were born, in good

[38] "La défaite du cancer" — Laurent Alexandre, JC Lattés Ed. 2014

[39] It was already known that accelerated male aging is due to mitochondria in the eggs and is interpreted as "the mother biblical curse".

[40] See: https://realinfos.wordpress.com/2012/07/09/naissance-de-30-bebes-genetiquement-modifies-aux-usa/.

health, in the US after a genetic mutation. Eggs were taken from non-fertile women and genetically modified with the contribution of the DNA from fertile ones. The corresponding babies originated, then, from three different parents together. After careful checking, the babies were found to be perfectly "normal" with good health. They do carry the heritage of the three parents and they would be able to transmit it to their descendents.

Evidently, this "experiment" opens the door to unimaginable variations; but we, yet, do not know if they will have the benefit of a life extension as mice did! We still have to wait no less than a hundred years!

There still remain open basic questions: for the civil status: who is the real mother to be registered? This makes it possible now to "mix" any woman with, say, a Nobel Prize woman winner or a famous star... or with a purely artificial DNA.

In the same order of the "tinkered" babies it has been recorded[41] that the first baby was recently born who was genetically "sorted" in a batch of embryos in order for him to benefit every guaranties of good health! You said Eugenism?

It then can be forecasted that the American sperm banks, so prosperous people say, would have soon to transform in simple industrial plants dedicated to the biological growth of cells, without any more donors, thus avoiding the present "tractability" problems.

[41] http://www.lepoint.fr/science/connor-premier-bebe-aux-genes-parfaits-30-07-2013-1709829_25.php.

Of course, the "father" (or, soon, the mother too) will no longer be black or white or Jude,[42] or Eskimo ... it will be "standard". It will be possible to choose between options as for a car. It will also become possible for aged or non-fertile couples to get descendants.

We may remember the moral revolution already raised by the "*In Vitro* Fertilization" (IVF); now are we ready to accept the "synthetic" baby? Then arises the fundamental question: is there an unacceptable transgression or an inescapable, natural, necessary evolution of the mankind.

Would natural procreation become avoidable? If puberty has to be delayed, will the natural aging mechanism be correspondingly slowed as in the mice case? There are questions we are to ask now. The different approaches of the Medically Assisted Procreation (MAP) or the surrogate motherhood are an area of severe controversies in the public space as well as in the governments.

---

[42] Jewish populations in the US are known to be especially careful with the selection criteria of the sperm donors.

Chapter 5

Brain and the Cognitive Science

Brain is, by far, the lesser known part of the body but which might prove unexpectedly exciting. Some 11 or 20 years ago we did not know much about the brain, apart from its anatomy and some electrical parameters. The available means of investigation were very restricted. The revolution came during the last years because of the explosion of new experimental means: computer and imaging. But the many discoveries are presenting more questions than bringing the answers. Anyway, the spectacular advancements every day bring plenty of news; each year databases double in volume and performances.

Each day affords its share of new, more precise and more complementary instruments. It is because of these instruments, essentially coming from electronics, opto-electronics and software that extraordinary progresses flourish in the knowledge of the brain and its "functioning". Will we reach a day when we will know everything about the brain? Some are saying that a full modeling of the brain is to be expected around 2020.[1]

Experimental approaches

The efforts recently accomplished in the domain of brain studies are substantial, the speed of result acquisition is fantastic as well as the emergence of new dedicated instruments which accumulate and complement each other. This tends to compensate for the lack of hindsight and the huge complexity of the question.

It must be worth recalling that, only 60 years ago, getting a simple X-ray image from the brain constituted a form of torture, and required complete anesthesia for a pitiful result.[2] The only remaining solution was to surgically open to see and derive a diagnostic.

Then, digital imaging arose which was recreated and revitalized by the computer and it was quite a miracle; it completely changed the story! The first element was the CAT (Computer Assisted Tomography), an X-ray scanner which made it possible to reconstruct images of "virtual slices", giving then a global and detailed vision in a three-dimensional and adjustable way. At the very beginning of the story, many, in France especially, including the professionals, were doubtful of the feasibility of an (initially

[1] Ray Kurzweil, The singularity is near, Penguin Group, 2005.

[2] I may personally bear witness of it.

English!) instrument which could compete, one day, with conventional radiography[3]; they absolutely lost the bet!

Some decades ago, just after 2K year, a tomographic X-ray scanner required 20 minutes to accumulate 600 images of brain slices and it was already a very revolution; today 3000 images are gathered in only 30 seconds! CATs are announced which will only require an irradiation dose 10 times lower than now; this is absolutely negligible!

It has then become possible to get images in real time and then perform a "functional" imaging that is to say quite an "internal video" to see the functioning of the interior of the organ in a three-dimensional way. It becomes possible to select the appropriate angle of sight, to keep free of the disturbing parts, zoom over a region of interest, and put colors in the different elements to make them more visible. But all these precious,[4] detailed and accessible information are poorly chemically specific, they mostly are morphological.

The extraordinary MRI

Then, was the MRI[5] which readily opened a larger field of observation. This method is an unexpected application of the nuclear physics theory: the Magnetic Nuclear Resonance, quite a physicist acrobatics supported by an engineering

---

[3]Objections were endless: resolution, X-ray irradiation dose, amount of data etc...

[4]The types of the tissues can be clearly identified.

[5] Magnetic Resonance Imaging.

miracle! It is absolutely harmless, unlike the X-rays; the apparatus is rather complex, using supra conductive coils at a temperature close to liquid Helium (4.22°K). They began to be routinely used in the hospitals in the eighties and never stop improving; they were very expensive at that time but they have become cheaper today.

The basic principle is to immerse the patient in a tunnel which provides a magnetic field intense enough to orient in the same direction as the magnetic moment[6] of the protons of the light atom nucleus, such as Hydrogen which is part of the water molecules included in numbers of organic cells. With the help of a radio electric wave it is then possible to make these atoms "waltz" around the magnetic axis out of the precession movement. Then the energy is measured when the atoms come back in a normal state; the local density of the atoms is then extrapolated and a map is elaborated by the computer.

The Hydrogen atom's relaxation takes place more or less easily depending on the tissue they are embedded in: bones, cerebrospinal fluid give rise to slow relaxations whereas greases or blood are faster. This gives a means to differentiate the tissues. Contrary to the X-rays, MRI is a means very well adapted to "soft" tissues. This way of analysis is especially rich in localized information. The spatial resolution is currently in the millimeter range.

In the particular case of the brain MRI, one is particularly looking at a visualization of the blood through the red cell

---

[6]One can also say the spin of the H nucleus.

hemoglobin which makes it possible to follow, in real time, the evolutions of the blood flow in the vessels, for example when a particular activity is stimulated. Then it is said that we are dealing with a "functional" analysis (MRIf). More recently, special models of MRIf have been especially developed for the research laboratories using smaller tunnels in order to get higher magnetic fields, then a higher resolution; these apparatuses are dedicated to laboratory animals such as rats or mice.

But the technology is boundless; one is trying to go much farther; researches are carried out[7] to invent a "nano-MRI" which uses a nanotube of Carbone as an antenna. One can get, by this means, internal MRI images from objects as small as a single cell (i.e., a neuron) then leading to innovative information on their metabolism.

Other instruments

Apart from the MRI, the field of the new instruments spreads from the macroscopic domain to the nano one; without a pause, other more classical instruments become more and more performing thanks to associated computers and software.

— Positon emission tomography (PET)
— Ultrasound
— Confocal optical microscopy associated with a digital image processor; this constituted a sensitive improvement

[7] In many laboratories including the one which (once!) had been mine.

of the classical microscope, allowing a three-dimensional discrimination of the structure of the tissue.

— Fluorescent markers which make possible spectroscopic investigations inside the cells or molecules with highly resolving optical fibers.
— Opto-genetics operated with an AFM microscope and near-field optics which allow manipulating genes inside the neuron.
— Electrical and radio-electrical methods to analyze signals from an electro-encephalogram with a computer. Reading the thoughts is already in practice in order to command robots or protheses.

All of this provides a wealth of new information at a macroscopic scale to investigate activated brain zones and as well at the microscopic scale to track down DNA in the neuron for instance.

It must be noted however that all these instruments get their information from outside. Obtaining a direct, intimate connection with the electrical or optical activity, obtaining an effective interface between the living and the inert matter are quite a different problem that is also investigated with patience and cleverness.

At the moment, micro-electrodes and other micro-circuits are constantly improved; they are required for the implants that began to be used, with success, in the heavy pathologies such as obsessive-compulsive disorders (OCD), Parkinson's disease, diabetes or obesity. Such implants are already of common use in the internal ear (cochlea).

Soluble electrodes or circuits are also experimented devices which spontaneously dissolve and disappear with humidity after a given delay or on request. Applications exist where the implant is provisional and must be removed through a surgical operation (pace makers and other captors). This solution also prevents the use of catheters and would surely be appreciated. At the moment, only the rats benefit from this technological advancement!

The magic solution to these technical problems will certainly appear some day and there will not remain any obstacle to the man/machine symbiosis.

The neuron, the brain and the body

The basic unit of our "grey matter" is of course the neuron, this mysterious cell, the powers of which are still unexplored; this cell, from a material point of view, sets out our so precious "self". Some 100 billion of them are actively working, bathing in the neural tissue which holds and feeds them.

The neuron

Yet unknown not so long ago, this cell began to yield its secrets and there are further exciting discoveries to come. Surprisingly, the first studies were dealing with neurons from squids, the simple reason was that those are the biggest!

The neuron was born 500 million years ago; it is a very complex and constitutionally diversified cell. There does not exist two identical neurons, each has its own specificity, its particular role. About 10,000 to 30,000 new neurons are created each day in the hippocampus or the olfactory bulb of a human adult.[8,9] So we presently know that these cells which were previously supposed to be definitely linked to us until our death in fact renew themselves on a regular basis, in small amounts.

Then, a "cleaning" mechanism does exist in our brain. Our brain is not restricted to aging. Studies are being conducted (always with the rats) to identify and stimulate the functioning in order to restore memory with "fresh" neurons and counter CVA, epilepsy, Alzheimer etc. ... (The rats will, then, be the first to be cured! Of course, they are known to be very intelligent animals!) One already knows that a continued cerebral activity also with a healthy and stress-free life does contribute to activate the cerebral renewing. The neuro-medicine is a recent science but so promising!

These new neurons, coming from an initial cellular division of stem cells, have a DNA molecule inside but, unlike all the other cells in our bodies which have the same hereditary structure, this neuron DNA is different, one from the others; we have just been told of it, and this for each new neuron which is to be created. In fact the starting point is just the same common DNA but the neuron also contains scattered parts of DNA and these parts can aggregate in a random way

---

[8]However, it is during the embryonary life that the growth is very fast: some 500,000 neurons in a minute!

[9]Contrary to the monkey's brain.

to the basic molecule thus creating a new structure. One says they are the "jumping" genes! Our neurons are fundamentally uncontrolled and natural GMOs. Our brain is a true "random GMO plant"! This is, at the brain level, an expression of the evolution laws to adapt the neuron to the variable requirements of the local environment and *stimuli*.

As soon as they are born, these neurons will move in the brain to get their place and their proper function, at the right place; they will transform, we do not yet know exactly why or how. They are mostly devoted to the learning and the saving of fresh knowledge. They will survive for a long time because they are known to die hard!

Neurons do contribute to the incredible network of interconnections, across the brain. A "good" neuron has the challenge to manage a minimum of 1000 connections (axons); if not it has "to commit suicide". In our chromosomes there is no definite wiring diagram but, instead, a perpetual activity for rearranging depending on the stress of the experiment (or learning). To think does modify the brain continuously. This brain plasticity has allowed Man to survive while adapting. Intelligence is not only a matter of genes but arises from this aptitude of the brain to react to the stimuli it receives. A "biology of awareness" begins to bring some light on these spontaneous mechanisms.

In 2008, in the Free University of Brussel one succeeded in growing neurons from stem cells and implanted them in the cortex of a mouse where they got perfectly integrated, which opened new perspectives to cure the Alzheimer disease for instance.

These possible modifications, these variabilities of the neuron give rise to changes in its behavior. The result of such a lottery[10] leads to the observation that two twins do not behave just the same. The new genes are not transmitted; they remain localized in the same neuron, giving it flexibility and easiness to adapt to the new role it must play; functions are differentiated and morphologies change with the function.[11] Moreover, synchronization is managed between the neurons through electrical or chemical messages which do not make the understanding simpler. Each neuron "family" (vision, motion, memory ...) has its own way of operating; links often operate over a long distance and, on top of that, they can reconfigure easily. The neurons benefit from an uncanny ability to work together, in synchrony, for absolutely unconscious operations.

This complexity, this flexibility, this diversity means that there cannot be two identical brains nor a permanent brain and this will seriously get more complex task of the software specialist who looks for a "mock up" of the brain.

It is now emphasized to find a mathematical and software representation which could reproduce some biological characteristics, especially of the dendrites, axon and synapses which are the organs the neuron uses to communicate: this could be called a "formal neuron". It has been implemented

---

[10]This lottery could likely be also at the origin of autism.

[11]One can selectively distinguish: perception, movement, emotion or purely cognitive function.

in a computer through different learning processes. Of course, the present models of the neuron are still rather simplistic (in spite of their mathematical complexity) but the advances are so fast! Actually, one does not try to reproduce the neuron itself but rather the neuronal function[12] which is exhibited.

IBM recently proposed a "cognitive chip" which uses a dedicated software structure called "neuronal network". The aim is then to coordinate these elements in a global and evolutive overall structure in order to learn and remember through experiences, therefore artificially recreate a kind of cerebral plasticity. The first tests are promising and belong to diversified domains such as associative memory, artificial vision, pattern recognition etc.

This has given rise to a new processor called "SyNAPSE" (5.4 billion transistors, 4096 cores in a massively parallel structure) which simulates a neurons network equivalent to a million biological neurons.

IBM also proposes a chip called "cat brain" which may look very presumptuous as we know the impenetrable intellect of these animals!

Another project is run in Manchester where Steve Fulbert[13] is trying to bring together a million elementary chips interconnected to provide an activity equivalent to 1% of the brain activity. It is not so much but, still yet, very much!

---

[12]Ray Kurzweil, *How to Create a Mind: The Secret of Human Thought Revealed*, Viking Ed., 2013.

[13]Steve Furber is at the origin of the ARM RISC — 32 bits architecture.

As for the Brain Mind Institute of the "Ecole Polytechnique de Lausanne", it plans to recreate, by a simulation on an IBM computer, the equivalent of a mammalian brain (of course a rat), neuron after neuron. This Brain Mind Project has been allocated a funding of a billion Euros and will use, in 2020, 1 exaflops[14] computer which still does not exist, and this, thus, calls for some reservations.[15]

Google, on its side, does not stay idle. It has begun to experiment an artificial brain which brings together 16,000 processors (a billion connections) to perform pattern recognition on human faces (and also ... cats!?).

The competition is then largely open to determine if "transcendent man" will be biological or mineral!

It is to be believed that the size of the human brain is equivalent to a bunch of 10^{18} bits; such a volume of transistors is estimated to amount to an only 1000 dollars in 2020. Furthermore, the computer memory would be a million time faster than the human equivalent! It will also rely on an infinite extension on the Internet network.

With such logical circuits especially designed, the human intelligence could be mimicked or even some "compartments" of it, as long as it would be possible to isolate individual functions.

When such a reliable computer's mock-up, a copycat even partial, of a typical human brain will be available, it could become possible to distinguish between what is going on well and what

---

[14]That is to say 10^{18} operations in a second.

[15]Michael Waldrop, *Computer modeling: Brain in a Box*, *Nature*, February 2012.

is not or what slows down and limits the process. Then, may be, there would be a move towards an "improved brain", more effective, more performing, more "genius" than the humane equivalent and which could perform tasks in the AI domain.

The science of the neuron is at its very beginning and is basically multidisciplinary, the domain to be explored is immense ... but research is making incredible advances. Here as in biology in general, everything depends on the new instrument, the technical serendipity ... also with so much work!

The brain in a whole

Another approach is to consider the brain as a whole, to understand how the information is circulating and how it leads to action, how it is stored and retrieved, how the process of consciousness works.[16]

Of course, consciousness (or, some say "cognitive scene") only occupies a very limited part of the total brain activity; it is only possible to think about a single thing at a time, but the underlying subconscious state is ready to rise; the unconscious state largely prevails, in terms of size of occupancy, not to mention the reflexes subcontracted to the spinal cord or the neuro-vegetative system which are more deeply installed, far away from the conscious state.

High level cerebral functions are located in the cortex with a high density of synapses: emotions, intuition, predictions, imitations, creativity, humor, etc. ...

---

[16] Le code de la conscience, Stanislas Dehaene, Odile Jacob Ed. 2014

From outside it is possible to detect an intense electrical activity in the different zones of the brain. The analysis and the interpretation of these complex signals are usually performed with powerful computers to decipher the sub-lying mechanism, decrypt them and possibly reformulate them differently. But there does exist simple user-friendly instruments that help trigger action by a mere thought.[17]

It is worth reminding that brain is the main energy consumer in the body, well beyond any muscle, even heart. Some functions of the brain are especially surprising and still not well understood, such as the vision which is provided through two independent channels and induces very fast pattern recognition that the brain is able to perform to recognize objects or detect a movement; response times are rather short (150 milliseconds) as long as a biological system is concerned.

Also, we are still unable to explain mathematically how to separate two sound messages which are simultaneously received. What are exactly the follow-up and identification criteria involved that are used by the brain to get these confusing messages separated? Physically, there is no means to identify each of these mixed channels. One does not know how it works, however, the brain does it routinely (to some extent), but not the computer!

The brain is likely to sort and identify the words at random, individually and in parallel and later reassemble them

---

[17]The *BodyWave* is a small box held on an arm and which detects the EEG signal prefiguring an action. John Cloud, *Thought control*, *Time*, November 21, 2011.

in the right direction as a function of own criteria (tone of voice, meaning of the words, etc. ...) in order to make a reconstitution of the supposed sentence. The large redundancy of the language helps to trace the meaning of the sentence of at least one of the messages. All of this operation is obviously perfectly unconscious but extremely fast.

Furthermore, we actually know that the occultation of the conscious state during sleep or anesthesia or during loss of consciousness, does not prevent the brain to continue its basic operations; software people would say it is a matter of "background mode" and this, the computer can comply with easily.

Sometime ago a woman, victim of an accident during diving, was in a persistent deep vegetative state; she was under MRI examination and it was observed that the brain correctly obeyed the solicitations (voice, suggestions, etc. ...) exactly as it does in a normal state.

Brain has a surprising "neuro-plasticity"; appropriate physical or mental exercises are able to generate a physical reconfiguration of the neuronal connections, thus creating or destroying synapses to induce a re-adaptation of the circuits to the new situation: the brain is able to "learn" and reconfigure. This is the way the brain recovers from a CVA or an accident. This would certainly raise some new questions to the software people.

Brain training leads to an optimization of its operation as a function of the type of problems and the frequency of the strains. This intends to provide a faster speed or a sharper analytic precision, for instance, in the examination of a face photography the brain is immediately able to analyze the facial features and identify, a similarity, an unusual detail. Present the same photo wrong side up and that is it, the brain becomes lost, it has to re-educate itself to get adapted to the new situation, gain a new habit; but yet the image the eye receives is intrinsically the same.

Now, how could we act on the brain? The knowledge already available allows to attempt to help with the use of dedicated implanted electronic micro-circuits.[18] Two technical obstacles, more or less overcome, still remain: the source of energy (rechargeable battery through an induction coil, with, in any case, a limited life time) and, more important, the electrical connection which is to be accepted by the living environment.

Micro electrodes are used, as, for instance, in the case of retinal implants which have been installed,[19] some years ago under the retina (node cells) of a blind man, in relation with the optical nerve.

Such a kind of prosthesis cannot be used for a blind from birth because his cortex has never been trained to the vision, but it is feasible with young babies who can still get this education. The vision process is an essential one which is to be acquired in the very early ages of the life when the baby has to learn how to see. His brain has to get trained to rearrange the disordered flashes coming from the retina photocells and create the feeling of a coherent image. This is in no way innate.

---

[18]They are called Brain Chips; 4000 persons are living currently with such brain implants.

[19]Institut de la Vision, Hôpital des Quinze vingt, Paris, May 4, 2011.

The body

The body plays an essential role in the brain activity; it is the natural interface with the external world. Its role is to record and transmit information to the brain, whereas the brain takes the required decisions and puts them to work. The body, therefore, cannot be separated from the brain functioning in the effective life process.

This is an important problem, leading laboratories are busy in analyzing how the brain actually operates and controls the deeds in good correlation with vision or touch, or with simple reflexes. The coordinated mechanisms involved in the muscles are largely multi-parametrical and the corresponding model difficult to carryout in order to obtain a "copy" a computer could use.

Broad outlines of the problem are beginning to be recognized and the "thought driven commands" are already in practice for specific and experimental applications. A project was to present, at the World Football Cup 2014, a couple of tetraplegia players equipped with exoskeletons[20] but at the last minute the project had to be changed for a simpler exhibition.

However, an example of a direct command through nervous influx still already exists with this artificial forearm fabricated by a British company which allows an amputee to perform the essential movements and "normally" use his artificial hand.

---

[20] Program engaged in collaboration with French CNRS, Japanese Riken, Duke University and MIT.

Such a detailed knowledge of the brain mechanisms is also needed to process effectively the brain function's re-education after a stroke.

So, from all of that, could an artificial brain be conceivable, should it be mandatorily associated to an artificial body? It is for certain that such a "system" would necessarily require a two way communication with the external world with the help of dedicated sensors. The intelligence, even artificial, cannot be exclusively academic, that is to say relying on the only memory. Even supposing that, in a first approach, the command of a "body" (whatever it could be made of) and the execution of physical actions are discarded to focus on the only "intellectual" functions, some "senses" will be absolutely required such as vision or hearing in order for the system to get a minimum updating self-autonomy.

The neurons in the limbs which are responsible for the movement are specific to a precise function; they each impose their orders: speed for one, direction for the other or amplitude as well. All of this is coordinated in real time with the vision, the hearing, the touch but the decision to act (it has just been discovered) is to be taken prior to our consciousness being aware of it, before the action takes place. The conditions for the movement, the coordination of the concerned muscles are programmed before the execution and sometimes corrected in due time. The unconsciousness directs us when we take our decisions and even in the action. That is the way a pianist can play a music score, and he does have to get trained for it.

Thought, intelligence and the brain

However, we did not wait until the computer and the MRI came to raise questions about our brains and the underlying subtle mechanisms of the thought.[21] No, said Socrates "Man, know thyself and you will know the Universe and the Gods?"

In addition of the neurologists and software specialists, now philosophers, psychologists, anthropologists of all kinds are trying to adapt their introspections with the new requirements arising from the scientific discoveries.

All these efforts are often grouped in a unique heading: the cognitive sciences or in short "cognitics", a very multidisciplinary field. An emphasis is put on understanding (one says now "model") perception, language, calculation, reasoning or even consciousness with the aim of making things accessible to the computer.

This operation is within what engineers call "reverse engineering" and it is intended to create knowledge engineering: the organization of the memory, the learning procedures, the mental representations are key elements of these studies because of their straightforward applications in computer science. We would not forget also David Ferrucci,[22] dealing with intuition: "I wonder what intuition could be, it is so named simply because no one is really aware of the process."

---

[21] Jean François Dortier, Le cerveau et la pensée, la révolution des sciences cognitives, Ed. Sciences Humaines, 2004.

[22] David Ferrucci is the designer of Watson, the IBM computer which won the game *Jeopardy*.

We already talked about Artificial Intelligence, what about simple intelligence? How to get a definition of it before trying to emulate it? There are many "compartments" and they communicate with one another. We are in the fuzzy domains of the psychologists, neurologists and even philosophers. May be we could make a rough distinction between pure cerebral activity (meditation, to some extent), and those which are, explicitly or not, related to the body and its connection with the environment.

In the first case, some forms may easily be put in relation with the computer, for instance, the logical mathematical reasoning, the classification of ideas, memorization organization. Others could be more difficult to be captured by the machine such as intuition, imagination, feelings which arise from the subconscious state with no apparent reason.

In the second case, the interaction with the body and the five senses are involved. It is nevertheless a matter of intelligence when keeping a relationship with the external world and in drawing the information to be processed.

In the other way, the cerebral speculations of any origins have to give rise to an action and here too intelligence is involved. This aspect will require the "humanoid" to dispose of the same (or better) sensors and also, especially, the "instructions for use" which are still to be elaborated.

To mimic the human intelligence (in a broad sense) with a machine is a somewhat vain challenge as this human intelligence is variable and very unequally allocated in the human species. If we need any reminder of that we need only to have a look on a TV program (any channel, any time!) to discover the limits of the human capabilities in the domain of intelligence!

It could be answered that television ratings do not follow intelligence requirements but more likely a more largely appreciated satisfaction for questionable amusements! It should then be better for the machine to be equipped with an artificial but not anthropomorphic intelligence!

Recent experiments

For a long time, it has been attempted to electrically interact with the brain. The Electro-Encephalo-Gram (or EEG) gave us electrical information on the working zones in the brain and it was imagined trying to stimulate them from outside to obtain a particular effect, either in a therapeutic intention or in the aim of triggering an active command, thus transforming the patient into a true "animal robot". Hopefully, most of these trials only concern laboratory animals. Some examples are as follows:

— Some years ago in a Chicago's laboratory a remotely guided robot was driven as a slave device by an artificially conserved brain, this "brain stem" was taken from, I do not know what primitive fish (an immature sea lamprey?) which is attracted by the light. Two photoelectrical detectors served as eyes and displacement dedicated motors were radio connected to the output nerves of the brain in such a way that the robot was able to follow the orders coming from the brain.

Of course, the experiment was computer assisted. The experiment worked perfectly well: each time a lamp was lit, the brain rotated the robot in that direction. It was also observed by the EEG that the reflexes get adapted to that point that if a sequence of light is repeated the brain understands and anticipates the movement. Would it be possible, one day, to make a human brain survive in a glass jar with nutrients and keep it connected to an artificial sensorial system? What would be the thoughts of this brain?!

In another lab, in San Diego this time, a lobster "*panulirus interruptus*" (why a lobster? Mystery of the science) was equipped with a cerebral electronic processor programmed in a "chaotic" logic to simulate a complementary biological system. The two entities, mineral and biological, have operated in a perfect symbiosis. No surprise for the lobster which gets perfectly adapted and shares its cerebral activity between the two sets.

The US Army currently experiments rats which are equipped with cerebral micro-chips and can be controlled at will in a labyrinth. The experiment works; the computer has been substituted to the animal willingness.

These simple (?) examples clearly show that interacting with the living is feasible. We can "put our fingers in our brain"! There is no reason why a human application would not work and this is not reassuring at all. Hopefully, there exist also positive counterparts which cannot be neglected in the treatment of heavy pathologies or accidental injuries and this is already on trial.

In this human domain, we had also mentioned the support for disabled which is in constant progress. An example of this is a paraplegic woman who is now able to activate a robot arm by her only thoughts thus allowing her to feed by herself. But the experiment can also be performed remotely: Recently a volunteer under MRI control in Israel was able to guide and make a humanoid robot walk in France. No need to say that military in every country is deeply interested in the offered opportunities. American drones in the Middle East can be flown from Colorado and the F 35 fighter could be flown by the thoughts thus preventing any movement for the pilot and giving him a better reactivity.

Recently, a hospital in Montpellier (France) launched a program in collaboration with Swiss, Germans and British with a working group of roboticians, doctors, psychiatrists, software specialists. The aim of this is quite extraordinary: to cure psychiatric diseases (autism, schizophrenia, social phobias) by means of a brainfull "re-education". We are in a situation of pure Eugenism but for a good cause.[23]

The principle of that is the following: a "virtual clone" of the patient is prepared by software modeling and simulation; that is, a "personage" who will look physically and intellectually as similar as possible to the patient. The patient will be trained to speak (?) and interact virtually with it up to reach a relation which could be qualified as "normal" (*sic.*).

---

[23]Keeping in mind, nevertheless that if the method works it could be applied same way for less good causes!

This clone allows the patient to be immersed in a virtual reality where he can see his clone, his *alter ego* encounter especially prepared situations in order to suggest a full identification with the robot.[24]

In the first stage, the simulation will be organized on a screen or a total immersion helmet but it is forecasted afterwards to make the simulation still more realistic with a real humanoid robot. (I am tempted to write "flesh and blood") with what the patient could talk and identify as a real self. Researchers are convinced that the suggested power will be such that it could become feasible at large dose and with an adapted pedagogy, to outright reformat the brain from scratch. Terrific!

There are no longer the fingers we put in the brain but likely the whole hand in order to reshape it without physically touching it.

Information, communication and the brain

Information, of every form, reaches the brain through the five senses of the body (namely vision and hearing); the various nervous links constitute the communication network. The returns are diversified, some can be fast and unconscious (reflexes) while others are slower and require some reflexion;

---

[24]It could be noted that our children or grand-children, in front of their play station already are submitted to a similar stress without a psychiatrist aside. One can forecast the psychological harm which is to occur in their young brains.

these returns can also just be a matter of memorization without any external action. The information the brain has been able to get, always requires an external stimulation.

In the old times the "brain work" was somewhat limited both in volume and speed, for the majority of humans. Communications were especially slow at the time. They were publicized by the announcements of the town criers or the homilies of the priests. These communications required of course a human contact and the use of a common language between the emitter and the recipient, with some care sometimes. Thus, Latin was used for a long time by religious preachers, doctors or lawyers precisely in order to restrict the information to only the person it was intended to reach.

Today things have changed, everybody has the right to information and rather requires it. It is an avalanche of continuous stimulations we are bound to. We even look for information when it becomes scarce, it is quite an addiction; sources of information have become numerous and abundant; the cerebral activity becomes frantic and nobody can get rid of it. It is as active and vital a principle as the air we are breathing. It is impossible to move without the 3G smartphone (awaiting for the 4G); impossible to get friends soon without being connected on Facebook; impossible to go anywhere if we run out of GPS! Some pretend that the level of a civilization is directly related to the volume of exchanged information.

New technologies are there to keep an oversized throughput rate for this information around the world to reach everybody. High frequency transistors, optical fibers, satellites are there to transmit all this digital paraphernalia: images, documents, sounds which continuously sweep over us. And this is not bound to cease.

The development of closer ties between man and computer is not already symbiotic but is beginning to be significant. Internet is there to provide an inexhaustible source of information but we still have to request them through approximate keywords and then sort the results. Google is not yet so subtle[25] as to distinguish the exact thoughts and the intentions of the partner; it analyzes the given indications but is only able to give back more or less adapted suggestions; finally, it is up to the partner to select the appropriate information. But this is just a beginning.

Advertisers already analyze[26] our relationships on the Web in order to decipher our personalities to target better pubs, more adapted to our own preferences, which makes it a real intrusion in our privacy. Will we still benefit, for a long time, from freedom to decide what we want to buy?

This process is then somewhat different from the way our memory is operating; it actually achieves a series of successive and fast back and forth exchanges and eventually, through approximation, reaches a satisfying answer (generally).

Google and others desperately try to identify more closely our wishes to enter into our thoughts, know our habits, by tracking our questions, and that is where the problem arises.

---

[25] A new version has just arisen which goes further in the analysis of the suggestions.

[26] Without asking us for our opinion and using automatic software tools.

Of course, at the very beginning, this intrusive curiosity claims at defending the good cause,[27] that is to say bring better assistance and afford the more adapted and fast possible responses.

But the final and perverse intentions also with the expected drifts cannot be ignored, for the time will come when the computer will know everything about our thoughts and our inner personality, and that is before we ask for any question. Social networks are already there to help.

But to reverse the roles, it just takes a step easy to make; Big Brother is there who can dispute the human at his free will.

All of that might happen without the machine being really integrated with the brain of the human; the action is already on the road to take place with all the mercantile and authoritarian hind thoughts we could guess. We do not dare imagine the new perspectives which such a physical intrusion, such a control of our brain through an implant, could bring in our life. We do not dare imagine the damages that a hacker could generate in a mob of men wearing cerebral implants.

Mathematics

This first part of the book cannot be completed without giving some words about mathematics. We would say that mathematics was, generally speaking, at the very origin of the Science in the ancient Greece and even before.

[27] However, Google, through its president Sergei Brin, presents itself as the "third hemisphere of our brain"!

Today this basic science is kept in the background of the "general public" news; no papers, no TV spots are devoted to mathematics because of the excessive esotericism and confidentiality of the science itself and the associated sophisticated vocabulary. Mathematics (at a high level, I mean) has become a purely intellectual domain which is reserved for highly educated specialists. This domain is exclusive to the point that Nobel Prize does not have a category for mathematics and so mathematicians were required to invent their own award: the Field's Medal!

This mathematical activity is purely individual and intellectual; that is the reason why I put this paragraph in the chapter devoted to the brain. Where should I have put it?

When the computer first arrived, the first instinct of the mathematicians was to joke: "This is a stupid machine which is limited to digital calculations of numbers; it will never be of any interest for us." And the story goes that the computers have been optimized to the point that, now, the software has brought up conceptual and formal problems; computers have become tremendously powerful; they can no longer be sneezed at, even by mathematicians.

In 1998, the Kepler's conjecture was demonstrated with the help of a computer. People are trying now to understand why and they are still looking for a formal demonstration so they developed a special software called "proof aid" to check for bugs in the program!

However, pure intellectual mathematics (needless?) is still alive, as shown by Andrew Wiles who, in 1998, succeeded in demonstrating the Fermat's theorem which was pending for two centuries ... without the help of any computer.

At the same time, the computer has also contributed in opening new fields in applied mathematics in various domains such as simulation, spatial research, meteorology, modelization, cryptography, or code theory. Mathematics still remains at the basis for any scientific development and applications.

Part Two

Hope or Despair?

Chapter 6

Some Considerations about a Possible Future

Given all these progresses in our scientific knowledge, given all these technological promises, what would we be able to reasonably extrapolate as for our mid-term future without entering the domain of utopia? Unquestionably, we are on the verge of a major transformation, but which one? What kind of planet are we to leave to our children? How will the politic be able to control anything, will it be swept along by the torrent?

In all this mess, we would have to integrate, on the one hand, the divergent points of views of the optimistic, who plans for the future without any restraint, with uncontrollable illusions, and on the other hand that of the pessimistic who waves the precautionary principle, the extreme caution, ideological prejudices, the antique values. Each one defends his own cause and comes with his own extreme activists.

Who is intended to do what in this concert, between the scientists who are supposed to know, the Beotians who are able to understand if it is explained[1] to them, and the broad masses of people who are affected without disposing of the means or the curiosity to try digest these progresses? The democratic rules of our societies would require that the decisions would be issued by everybody (Is this reasonable? Pythagoras was of the opinion that it is foolish!); but in practice, one can observe that referendums are very rare, because the governments, whatever they are, precisely want to ... govern! The best scientific intentions can be sometime wrecked by an inappropriate or clumsy use that is to be imposed to them.

Caution and risk-taking must be taken in balance, but the wisdom of the people is uncertain, even more so that we will consider here only the extrapolations which are conceivable starting from our present knowledge.

Many other new opportunities will certainly arise tomorrow, about which we have no idea today.

What questions could be reasonably raised?

The advancement of sciences leads to question ourselves in several directions: first of all, there are questions which concern Man exclusively: health, death, survival in a changing environment. Then comes the material context, the role the

[1] You will always find some "intellectuals" concerned about ethics, to fabricate evidences of an opinion which cannot be challenged.

machines are to play, the always stronger role of the robots. There also is the human society itself which requires to respect the accepted rules, if a global coherence is to be preserved.

One thing should be taken for certain: the steamroller of the Science comes right behind us. Inexorably it brings us, every day, new opportunities and requires answers and decisions. If we do not want to be crushed, we should decide what to do and say it quickly.

There will certainly be bad choices, errors made along the way, accumulated harmful confusions; the laws of evolution which were there up to now to guide mercilessly our destinies are going to disappear precisely for our will to take charge of them.

We should have to be collectively vigilant and rational, but things will happen as they have to do, ineluctably, for our benefit or against us.

For the individual man, the basic question remains: would I be able to cure my illnesses and pains, would I be able to live longer? Current progresses suggest that solutions, opportunities, eventualities are within our reach, without knowing precisely which could be the more adapted neither in what order they might occur.

Beyond this survival which is supposed to be biologically managed, the question arises to go farther; will it be possible to "improve" the human, to provide him with enhanced capabilities to think, act and live differently? Yes, new possibilities are there with all kinds of prosthesis, with the assistance of an advanced intelligence, with an organized communication. But for, what to do? For how to live: to work, to have fun, to make war or to have Mars colonized? What fun could we expect, individually, from this ultimate mechanization? Is fun still valid?

One can be skeptical, considering the way the modern life engages our days by disposing our time, so preconditioning our leisure from birth to retirement (and farther!).

Collectively, questions also abound about how our societies could change. Currently, they are in a constant turmoil; they search for themselves: How to bring this humanity to live together, peacefully, to endure each other?

Pooling the problems with the new communication facilities, the globalization of the relationships, the mixing of populations, the sharing of the knowledge, all of that means that a global organization will become mandatory. Would it be authoritarian, libertarian or programmed, optimized and computer managed? Would it be possible to remain for a long time to do just anything, regardless how, without an accepted ruling?

What the robots and the computers will be able to bring to us?

The reader would certainly have noted the many references to Google, this Californian company, which got the dimensions and the power of a real state in the state and which interferes in every kind of domains. It is now time to deal more extensively with it.

Big Brother or Big Google?

In the galaxy of the companies which support the scientific and technological progress, a new star was born some years ago;

since then, it has become especially bright and voraciously feeds off all what crosses its path or which could carry a juicy project. This star (some say "this black hole!") is named Google.[2]

Similar to big companies in software and electronics (Apple, Microsoft, Facebook and so on), Google was born in a garage in the Silicon Valley from the imagination of two visionary students fresh out of the University of Stanford and firmly committed to conquer the world: Larry Page and Sergey Brin. The times (1998–2000, not so far from now) were then to the advent of a network connected between the American Universities, the military and some big companies. Increasing volumes of information exchanged and stored in the computers make it necessary to implement a regulation (Internet) and fast and efficient means of accessing and sorting in order to exploit this unexpected manna. New software appeared which were dedicated to this "search and found" operation: the "search engines".

Everyone had his own recipe and many start-ups blossomed, attracted by this exciting opportunity: Yahoo!, Lycos, AltaVista ... then Google arrived and made its mark as a leader, seriously cleansing the competition all around it.

Google strategy

Different to the previous competitors, the newcomer right away understood that if a search engine has to be a freely avail-

[2]Google, un ami qui ne vous veut pas que du bien, Pascal Perri, Anne Carrière Ed., 2013.

able to the user to make his researches, nevertheless things do not stop there. It is also a powerful advertising medium which is potentially lucrative, and, more of it, provides an ongoing source of valuable information ... to the user himself precisely.

Then Google started gathering (with total indiscretion) all kinds of possible personal information on everybody, thus allowing to better target returns on advertisements. This data bank, of course, is very well kept that no unauthorized individual could access. This is the "war treasure" of Google. Formally, of course, this is done in the generous aim to offer the best services as possible (always free, as nobody can resist to the "free") to the naïve consumer.

But "Big Google" has the means to control everything. The "services" of Google are also growing in numbers and diversity, all converging towards a finer investigation and a "siphoning off" of information: Gmail reads and stores all the messages (sorted and classified), Google+ scrutinizes our relationships over its social network, Google-Chrome analyzes our researches on the net (particularly the blank pages!), Android, free operating system offered to developers, typical Trojan horse, now operating 90% of the smartphones, consciously spies conversations and research subjects on the Net, Google-Now virtual intelligent aid installed on Android listens to our wishes and harvests our daily acts and gestures, Google Books is the monopolistic referrer of the editors,[3] etc. ... The takeover of the customers

[3]Editors then became the victims of being more voracious!

is now complete to achieve a worldwide "neuro-marketing" and more especially on the flourishing European market (and more particularly French because French consumer is known to be especially naïve, dim-witted and vulnerable). Obviously, this formidable weapon of the "free stuff" is complemented by a locked monopole. Everything is done in a total lack of transparency; as Google is deeply inquisitive on everything as obscure it is on its plans to use all the information it is harvesting.

The Google's search engine has become a quasi-monopole in Europe but its functioning is kept absolutely secret; it is not possible to know the criteria of the referencing which is practiced by the "Page Rank" algorithm. The leading rules get permanently changed in order for Google to remain the Game Master. No announcer will be able to use the rules in his favor and so improve its score, that is to say, its accessibility.

Google, through the power of the acquired intelligence, can already eliminate or promote, at its option, one advertiser, as a function of the usefulness it could assign to it. So Google can, with impunity, put in practice the centralization and the crossing of files and nothing can be done against it. Google is now attacking the market of the "Connected TV" and so it is going to introduce itself into every home in order to daily monitor what we are watching on the TV. All of that is performed in all innocence, apparently. The Google profession of faith, which is widely repeated, still remains "do not be evil"!

Google, an architect for a new order

But Google's ambitions have not stopped there, Google has become a technological octopus, cannibal and voracious,[4] a machine for "global domination"[5] which intends to be the engine of the "trans-humanism". Ray Kurzweil, the guru of this philosophy is now the boss of the futuristic innovation at Google and there is no coincidence!

Google is everywhere a scientific or technical novelty could appear, buying or controlling the corresponding companies. Google has used successfully its strategies and currently represents a first rank financial power in the world, which allows every fantasy. Already in the past Google had taken control, in many domains, over key companies of the business like: YouTube, Aol, Picasa, Boston Dynamics (robotics), Deep Mind (Artificial Intelligence) etc. ... Google is the owner of a million computer servers throughout many "storage plants" all over the world ("cloud computing").

Google is everywhere, even in domains where it was least expected: Medicine, Robotics, Artificial Intelligence, Data analysis, Cars, Mapping, Edition etc. ... The political project of the giant covers everything at a world scale and Google peacefully announces that its project is to "digitalize the Universe", nothing less!

---

[4]Le Monde, August 28, 2012.

[5]Pascal Perri, *op. cit.*

A tour around this hegemonic empire

— Medicine is an essential sector of the future which is to dominate our private life, promising to help us for a longer life. The just raised ambitious program "Calico"[6] intends, in the mid-term, to reduce morbidity by targeting the aging process and degenerative diseases (namely, Alzheimer disease). Google forecasts that medicine is to become an information science and therefore has become an ideal target for Google which has the ability to gather and process enormous amounts of data[7] with the aim of finding corresponding diagnostics and therapies.

— Google currently talks about "mastering the brain" with neuro-prosthesis (?) driven by Artificial Intelligence (Ray Kurzweil) which could keep an open door to a neuro-dictature.

— Genetics of course belongs to the considered technologies. Google's CEO, Larry Page announces quietly "one day that might defeat death itself!" Lung cancer, after the same Larry Page, would be overcome by this new approach, but that would not contribute significantly to notably increasing our life span, whereas genetics would. Other giants such as IBM or Oracle are also on the track.

— Robotics and AI, both together, are targets for Google which has the necessary know-how and can afford to buy

---

[6] *California Life Company* (Calico), project directed by Arthur Levinson (CEO of Genentech).

[7] The new digital Age, Eric Schmidt, Jared Cohen, AA; Knopf Ed., 2013.

the most experienced companies in the field. Some achievements can already be mentioned:

* *Big Dog* is a military four legged robot able to move by itself in the worst difficult terrains, smoothly rising in the event of a fall, all with a 200 kg load on its back, and in total autonomy. But the last point is the Achilles heel of the *Dog*: the battery. The more the efforts, the more the electrical consumption!

* *Google car* is for real with all corresponding derivatives such as the truck's convoys or the automatized taxis (or as well, planes or drones!).

* Google glasses which bring man in permanent relation with the virtual world.

* *Mapping 3D*,[8] an application freely accessible for the Android smartphones which is also freely given to developers. As a matter of fact, these gifts do have a downside: when further applications are developed they all are at the mercy of the base software, that is to say, Google!

* *Google Earth* and *Google maps* are going with you in any of your trips. (This is now on the way to be extended to the submarine domain!)

The list is long and non-exhaustive. It is still constantly enhanced with new futuristic openings, namely the software of the future. Google is ready to take all the bets. Quantum computer is still in limbo but this did not prevent Google to

[8] Android 3D mapping smartphone, Google's Project Tango.

deeply invest in D-Wave. The first quantum computer was consequently built up and immediately bought by Google to assess the possibilities even if nobody presently knows how it works or if only it works or what could be done with! Nothing will escape Google! If it does work, then Google will get the monopole of this crazy innovation.

Is Google really almighty[9]*?*

Millions of machines, a monstrous cash, the best scientists who are well-paid and supported, everything's ready and locked to provide an absolute power. A single slogan is put forward: "Get the moon!",[10] which means "go ahead!" The intent to set up the future world empire is clearly stated and put on the way with lucidity, tenacity and caution. An ethics committee has just been created in order to define the limit of what could be accepted by the present society and what could be the way to get around: How far to go, right now or later?

Google, until then, succeeded in withstanding all destabilizations attempts. Now, it is able to hire the best, more wily lawyers. Google has recently been sued on the charge of the liberty it took to reference as it sees, on its search engine. Then, it succeeded to make the American justice accept that the machines (GPS, YouTube, Facebook and so on) are "real

---

[9] Pascal Perri, *op. cit.*

[10] To be taken literally as well as figuratively: Google is also interested in Moon travels!

speakers" as well as a human and then deserve a law guaranteed freedom of expression! What a way to go!

However, there is even more actions to court, in Europe (fiscal laws and CNIL[11]) as well as in the US, without any success at the moment. But the move is on for a resistance. Google will not be able to develop indefinitely as a state in the state. The absolute weapon that still remains is the only "anti-trust" American law which might allow only the President of the US to cut the giant into several independent pieces in such way that a global strategy could no longer hold or be synchronized and that the financial power could be split up.

Changes are already on the way

Our social life is already deeply affected, transgressions are there and they have been integrated in our way of life, in a constant change; among young people of today, the family does not exist anymore, the "mobile" imposes its virtual reality, juvenile delinquency is on the rise, marriage has become accessory (excepted for gays), abortion remains a badly supported constraint, single parent or single sex families abound, parents are busy to "fulfill themselves professionally" sometimes on the far side of the world, children are grown in a nursery and then fed in a canteen without any maternal proximity, finally they become a discomfort for the prosperity of the parents, mainly when grand-parents are still not there to

[11]Commission Nationale Informatique et Libertés.

bridge the gap. Familial life cannot allow for the "career mobility" the modern life requires in our mad society.

Some elements which may bring a change are already observable around us. They are of individual order but also collective, social, behavioral. They were assuredly quite unthinkable just some years ago but they give rise today to problems not so easy to solve. Some could be mentioned as examples.

* Gays and their frenzy of marriage[12] constitute a recent social epiphenomenon; not that gays did not exist previously, they assuredly do exist from all eternity, without being a problem for the society, nor the government. Is there really an epiphenomenon or rather a prescience of a coming disruption in our sexuality habits and breeding which in a near future will no longer require bringing together two partners of the opposite sex? Such curious comments come from Alain de Benoist[13]: "biological sex being no longer determining, each individual can freely determine his (her?) sexual identity ... It follows that everybody is able, from his own drives or mood to freely build his sexual identity, or change, each construction being a norm equal to anyone else." But Simone de Beauvoir, a long time ago, still warned us "one is not born a woman, one

---

[12] Marriage is of interest for only a minority of them; marriage is the only pretext for social destabilization.

[13] Alain de Benoist, Eléments pour la civilisation européenne, October–December 2012.

becomes one". Such nonsense will have their place in our lives in which politics and sciences do jointly contribute.

* This is all in accordance with the "gender" theory initiated by the psycho-sociologist New Zealander John Money[14] who refused the biological sexual differentiation, in favor of psychological, social or purely behavioral criteria, the notion of sex being not considered as innate learned or even imposed by the society.

This subversive theory[15] today develops in a totalitarian form which wants to lead to a society where the imposed natural sexuality will be removed at the same time as the biological family notion. The French Minister Vincent Peillon publicly stated: "The aim of the secular moral is to extract the pupil out of any familial, ethnical, social or intellectual determinism". This situation implies that the religious question has been withdrawn, which is not straightforward nowadays because of the violent opposition of the Muslim populations.

Then, there is no need for the Science to push us towards catastrophes, our "intellectuals" surely are capable to invent, by themselves, our own transgressions and impose them.

* The availability of all the means of information has led to an inflation of communication we have never seen before.

---

[14]J Money followed famous Dr. Kinsey.

[15]Gender theory is already taught in the universities, high schools and even elementary schools. This theory was put into practice in northern countries, but recently Norway finally gave up.

The teens are living in a world of SMS, video and virtual games we do not yet understand fully and in which they are alone to live by their wits, without any guidance, rules or ... spelling! The notion of morality has been excluded from the vocabulary. This adds to the loss of the family framework, the disempowerment of the parents, and the impoverishment of the education system which is intended not to share high quality knowledge[16] but rather to summarily deal with a transitory important stream. Some argued that: "the fathers disappeared and the children became little monsters, capricious, soft and tyrannical."[17] This is not entirely wrong and goes with a disturbing rise of youth crime and vandalism. All of this makes an unwanted and unscheduled outcome of technologically poorly integrated progresses.

* To this, we can add the sad picture of the youth's addiction to the virtual (and even reinforced) reality of the electronic games and systems the realism of which is more and more astonishing. Nothing is to surprise, everything is allowed, everything is accessible.

From passive pupae, respectful of the adults, waiting to grow, children have today become active butterflies, independent, active with an ability to take decisions, fully informed from very young age, but, dominantly, without any insights, they remain inexperienced and anchorless, with no culture.

---

[16] Latin is almost no longer been taught whereas Greek studies disappeared.
[17] Dominique Venner, *La Nouvelle Revue d'Histoire*, July–August 2008.

More dramatic, through Internet and the ubiquitous iPhone, youth has become very sensitive to the new world which is offered to them under the cover of liberty, openness, and self-discovery. This arises precisely at the moment when teenagers begin to rely more on their peers for support and less on their parents.

Addictive sites of social networks (or more likely anti-social as Ask.fm, Secret, Whisper or YikYak) especially dedicated to their intercommunication, have appeared which are real psychological traps. Members of the site can publicly ask questions and post answers anonymously. They have been set at the time when the fragile psyche of the teens is booming.[18] The teenage brain is a confounding organ where differentiating between good and evil is not yet achieved.

Some teenagers are driven by vicious messages to commit suicide. The organizers of the sites give a single watchword: "We teach people to bully." An outright ban shall be of no use. Close down the site, you will get another one soon.

How would such young minds get rid of this underhanded grip, how would they segregate the real from the virtual? Does this announce a new way of life yet to be invented? Be as it may, this notwithstanding, materiality is always here.

We are ripe for a deep mutation of our very existence. But what existence? May be a new species, a *"Homo modernicus"* (or *"Homo-Googlus"*) is in the making, which could succeed *"Homo sapiens"* Cro-Magnon as this one replaced *Neanderthals*?

---

[18]Recent findings in neuro-sciences have shown how the developing brain is ill equipped to override emotional reactions.

Convergences and divergences towards singularities

Ineluctably, this multitude, this diversity, of new approaches require, each on its side, a deep call questioning our being; the new man born out of this change will enjoy unthinkable benefits, but he will also have to pay the price with obligations, constraints, drop-outs, downsides which we are not yet aware of and which will possibly not be pleasant at all to live with (for our generation at least).

Today forecasters ensure that a "Singularity" is to come before we could reach the "trans-human" ideal state of long life. Instead one could rather infer that many successive singularities are to come, piled up one after the other, each one corresponding to a particular "improvement". This could happen in a rather short period but the "post-human" man will certainly not rise all of a sudden as Minerva from the skull of Jupiter, helmeted and armed. Many "prototypes" will be necessary, also with ends of series, beta versions, some failures too. They must be different and competing models. Singularities will arise from all directions and will surprise us.

Then the "wall" will be multiple, stratified, with uneven layers which we will cross one at a time, and bouncing. Each advance in the Science will result in an improvement of the following model. Will we reach a final and definitive realization for the centuries to come? Nothing could be less certain; we now have to guide our destiny and Mother Nature is no longer involved in. We are doomed to manage ourselves and this will require a good deal of rationality[19] not to be drawn in a nightmarish universe. With this new frontier on the horizon, we would have to take into account the economic interests that some "carpet baggers" certainly would like to make fruitful regardless of the justifications.[20]

In this connection, we could mention the "panga" farming and industrialization in the countries of South West Asia, namely Vietnam on the banks of the Mekong River. This omnivorous fish grows very fast and it is raised in frightening conditions of insalubrity. It is obtained in large quantities by using artificial ovulation through Chinese hormones secreted with the urine of pregnant women. The cost price is very low and because of that this fish is integrated, free of advertising, in many by-products sold worldwide by a large-scale distribution. If it would be possible tomorrow to further accelerate the process with a gene mutation, no doubt this will be done, inconspicuously.

The confidence we could pay to men would certainly have not to be wrapped with naivety.

Lengthen, duplicate, and synthesize life?

There is an ancestral competition between the population growth and the depletion of resources; a number of episodes happened from eternity where devastating famines were fol-

---

[19]About the legitimacy of the scientific talk, see: www.youtube.com/watch?v=uORDvWwbE_E&hd=1.

[20]This will certainly be more tricky than the management of the shale gas!

lowed by fast recoveries. Malthus was the first to state the problem, until local, in its planetary dimension, but agricultural machinery came just in time and changed the equation. This opportunity calmed down the immediate needs but at the same time it has restored the growth of the population.

Today the ever-present problem of the "lack" remains the same except that it has considerably broadened: our appetite for consumer goods means that it is no longer a matter of food but more generally all the resources required by our modern way of life, more and more demanding, more and more invasive. It is not just about oil and gas or corn or bluefin tuna, but also lithium, essential to the electric cars or "rare earths" (namely, the lanthanide family) without which there are no possible integrated circuits, and no smartphones. Of course, it can be said that a new technology drives out another but the basic needs still remain the same.

Today, biologists promise to bring us immortality in a quite near future. This tendency towards immortality is to progress regularly and is undoubtedly going to augment the population. The evolution mechanisms will be blocked thus excluding us from the natural laws which up to now took care of us.

Before getting to that point we would have to combat diseases we do not know yet. To maintain our body alive should require an early prevention of the illnesses but also a permanent monitoring, even personalized and frequent cares. In such a context the machine will certainly be of a necessary help.

The need has already appeared for a global management of healthcare through super computers getting together all the medical knowledge (mass-customizing the drugs) and also for tracking personal information files. As such, it is not surprising[21] that IBM be in the running; its famous parallel structured computer "Watson" has been converted to afford this new experimental approach: a doctor gives the name of the patient, the corresponding biological analysis and Watson gives back the information also with the diagnosis and proposes the corresponding medical prescriptions. At this time, it leaves it to the doctor to judge before acting!

An organism which does not die, no longer has the need to reproduce but accidents still happen and thus the species could disappear if it does not reproduce enough. Death is necessary to the survival of the species. If humans become immortal a subtle mechanism would have to be invented to make life conveniently regulated. Would this mechanism apply to everybody or would it have to select "survivors", those who enjoy necessary predispositions, those who are worthy of help, or those who can afford the service?

A possibility also arises that this over-population of aged people be complemented by newcomers: the children who could result from human duplication (cloning) or artificial creatures (we do not dare to say "children") from genetic programming.

Everything is keeping with an uncontrolled increase of the world population and that is where the shoe pinches! In any

---

[21] Fortune, The supercomputer will see you now, December 3, 2012.

case and without waiting for biological miracles, we have already entered a critical stage of over-population which is to impose, in the mid-term, drastic decisions. At the time the balance is always on the side of a constant growing of the world population. This is assuredly a major component in favor of a near "singularity": we can see on the graph, here below, the evolution of the world's population (we have cheerfully crossed the line of the 7th billion and some applauded!).

I tried, in vain, to fit the population data with an exponential model: it is worst; the closer approach seems to be hyperbolic with an asymptote, that is to say a catastrophic runaway towards infinity; this indicates an inevitable "explosion" before the year 2034 (roughly)!

Well! There are forecasters who do not share this position; they hold the promise that a spontaneous trend reversal is to occur soon, the reasons of which are diverse and multiple. However, not any sign of slowing can still confirm these optimistic elucubrations. The more this slow down would last, the more it will be of violent consequences.

It is to be feared that this would generate a deep transformation of the age pyramid; then we are on a delicate balance.

This is a problem right now in Japan where people are very serious to find technical solutions (robotized domestic help) that could be imagined to be taking care of an aging population.

China also, in conjunction with its recent prosperity, is facing a decline in the birth rate and has begun to experience unexpected difficulties. Some 25 years ago, China opened

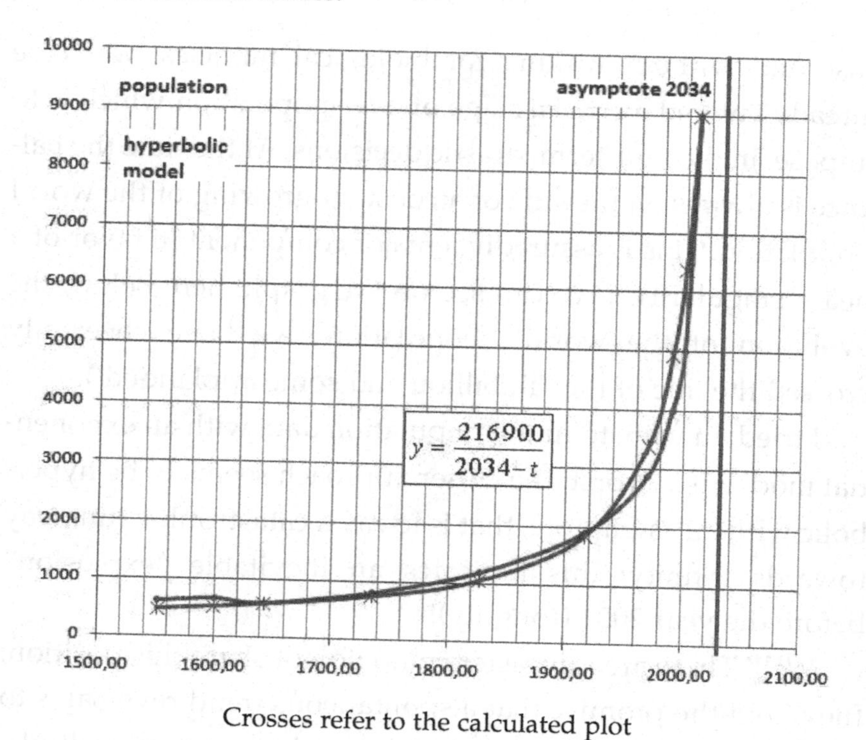

Crosses refer to the calculated plot

international relationship; the country was rather poor, the infrastructures were non-existent, the natural resources badly exploited. Mao promised a bowl of rice for every Chinese; but to live up with that promise there were two solutions: either increase the number of rice bowls or lower the number of Chinese. The second solution has been put up with the draconian policy of the "only one child". A severe blow was given to the demographic growth.

During the following years, China laboriously got enriched with its foreign trade and was established as an undeniable world power. An important class of urban petty bourgeoisie appeared and madly jumped into the consumer driven society (homes, cars, every kinds of equipment, university studies, etc.) turning its back to the previous agricultural society.

This new prosperity in a phased manner forgot the "only one child" politics. One would have thought that birth rate would increase significantly to fill up the hole left in the bottom of the top-shaped scheme of the diagram (see below). But nothing came and the situation has become worst: young families without children have become the rule and the hole is getting bigger instead of filling up.

What did happen? Having a child in China has become a luxury that only few families can afford. Social welfare does not exist, two salaries are necessary to make the living, studies require payment of fees, higher cost of living requiring expensive modern equipment ... all these negative effects accumulated; today, children are scarce and perceived as an inaccessible (and unwanted) social valuation. Economic growth has been very fast and the social evolution has got out of hand. Chinese acquire the needs of the modern life without having, yet, the means for it. The effect achieved is the exact opposite of what is desirable.

A rapid slow-down in the Chinese demography is expected to lead to a shortage in the industrial manpower required to keep pace with economic development. This along with the rise in the salaries would result in a destructive syndrome.

If this example is to extend to other developing countries, it could lead to an inversion of the present growth tendency of the world population which would prevent an explosion but at the same time also generate new dangers to be mastered.

Growth is not always eternal, as well as the globalization which up to now has been considered as an unavoidable[22] guaranteed future.

Elsewhere, as in Africa, there is the reverse situation of finding adults able to take care of an excessive and unattended young population.

In a very succinct word, the biologist Aubrey de Grey[23] warns us that, if our longevity is continuing to grow a drastic choice will have to be done between "not die and have children".

Indeed the problem is not so sketchy. The situation is quite different from one continent to another, from one country to another and even in a given society from one social class to the other. A global reasoning would lead to noticeable evaluation errors. The ensuing transformations will not reach everybody at the same time and the corresponding amplitudes would certainly not be the same everywhere. Also, the possible solutions would not be the same everywhere. Then a relativistic and smoothed vision is certainly to be preferred.

Age pyramid and longevity

The age "pyramid" (see schemes below) shows the distribution of the population density as a function of the age. The

[22] American economists already observed a reduction in the globalization in favor of the domestic markets which look safer.

[23] Aubrey de Grey, Michael Rae, *Ending Aging*, 2007.

data here are only indicative. Of course, this diagram is continuously evolving.

The first scheme (a) refers to statistics available in the 19th century when populations were rather stable. It is observed that the envelope strongly bends from the bottom, it struggles to take off the axis; this to be ascribed to the still important child mortality. The curve largely unfolds to reach a sharp tip in the century range, centenarians are rare (almost no men).

The second scheme (b) shows the situation at the end of the 20th century in France. The situation is not stable but rather obeys a dynamic, transitory state. One could note that the triangular shape is no longer followed and has transformed into a round form. Today the child mortality is much reduced (the curve takes off abruptly from the axis and the longevity sensibly rises); the tip is larger and the proportion of centenarians gets higher. Such an evolution is quite normal owing to the explosion of the "life expectancy" which pushes the curve to the top.

Today there are more than 17,000 centenarians (men and women) in France, from only a hundred in 1900; it is forecast that (with constant living conditions) there will be 75,000 in 2030! The pyramid takes a bellied shape and the proportion of "actives" (between 20 and 50) rises before likely diminishing.

The last scheme (c) depicts a situation called the "spinning top"; it shows what would be the situation if the births were to be severely reduced thus choking the curve at the start. Of course, this is a typically transitory situation which is not to hold for a long time. An inversion of population is to

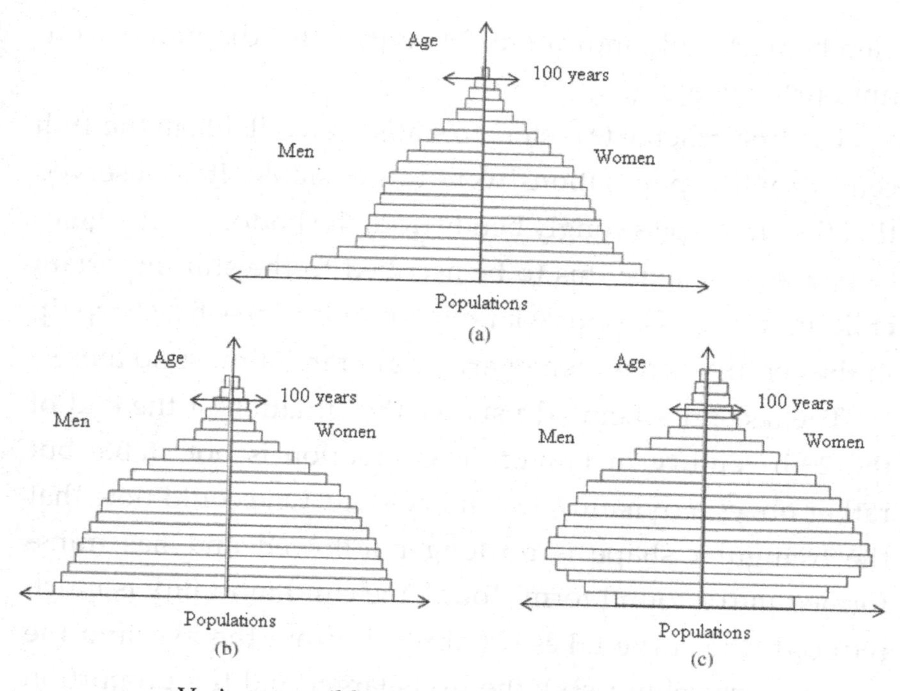

Various possible situations of the age pyramid

occur; young people would not be numerous enough to provide a generational renewal and the sustainability of the retirement pensions.

Such a decrease in birth rates could arise from two reasons: the increasing number of abortions which deprives us of more than 200,000 babies (in France) and the observed declining fertility of our young[24] males. This worrying situation is overshadowed by the entry of exogenous populations,

[24] A possible explanation of that could be some chemical molecules found in the drinks, migrating from the plastics of the bottles. Another contribution, which is attested, is to be found in the cannabis use.

young, prolific, without qualifications and poorly adapted to modernity.

This new transitory top shaped form is worrying and presages the possibility of catastrophes, the active population not being able to help sustain livelihood. It still remains that appropriate responses are to be found in the robot assistance (Japanese solution) or, above all, the "recycling" of the seniors in order to keep them active for a longer time. This will be necessary more so as the life expectancy keeps rising.

In all these age distributions we discussed, women seem largely preferred in numbers (for an equal risk weight), but not so in life quality. Nature behaves strangely when gender is concerned: male children are more numerous than girls but at the same time they die more frequently before puberty is reached; this restores the balance which flips again in favor of the girls at the end of the life.

Men centenarians only represent today 14% of the total population, which is quite weak a percentage. This proportion could rise in the future when the women's work will enter the statistic of the age range, not that women were inactive before but many did not undergo the professional stress which they are currently subjected to, since they got "freed".

This extended life expectancy comes from the compelling improvements in the standards of living (hygiene, food, medicine, vaccination, comfort, sport, etc.). If we need any reminder for that we need only to review some photos taken a century ago: miners, workers, urban or in the field activities. Particularly, the poor appearance of the clothes is really scary. It provides a sense of dismay and this question: "How did they overcome such life conditions." Such a living is not conceivable today as we have got out of the habit of damming the socks. Would we have to feel superior, or more lucky in life?

But a second question arises: Let us try to plan the future; a century from now; what would our descendants think by seeing the pictures of today? Would they feel a degree of commiseration our present primitive way of life? What will they be, themselves and would they dare judge us?

We also must not forget this evidence: our present centenarians ... were born a century ago (that is trivial) and, when they were young, they had to survive a situation unparalleled with the present comfort.

Within a few years, there came the "baby boomers", this wealthy post-war generation of the "Glorious Thirties" (years 50/80). From the very beginning, this generation had benefited of sanitary conditions previously unknown: penicillin, antibiotics, vaccines, social welfare facilitating healthcare, etc. During this period, unemployment did not exist, the annual number of deaths in France broke an unprecedented record dropping from 145,000 to 25,000 a year[25] as a result of a drastic reduction of the child mortality already diminishing before WWII and also because of extensive improvements in the healthcare (tuberculosis, smallpox were eradicated and

[25]This is despite a notorious increase in the population and a war in Algeria. These events did not have any statistically noticeable influence.

even cancer was more efficiently cured). Many more people are to reach the high age.

Their successors who were born end of the 20th century will have benefited from a still more improved sanitary context, no doubt that there will be many born during this time who will cross the line of 100 years. It can then be forecast without fear of contradiction that a solid generation of centenarians is to happen soon.

Also one could say that there have been two world wars in the 20th century[26] and wars are known to vigorously stimulate technical creativity. Would we need a third world war in the 21st century to stimulate us even more? The answer is: likely not, because our activity is already largely promoted by the increasing need and wild economical competition.

* How to live longer today?

Recipes are well known, even if they are not always well fulfilled. Everybody has its own recipe, of course, but there are key guidelines to meet. Physical exercise is mandatory to keep the biological[27] machine in order but it is also required to maintain the mental machine: always keep an ongoing intellectual project and an optimistic moral; all of this is necessary to ensure a proper resilience to the aging. To be curious about others and foster a certain philosophy (with also, may be, a touch of selfishness or narcissism) can help survive.

[26] Including 40 years of technically stimulating Cold War.

[27] Obese centenarians do not exist.

Centenarian men, even if less numerous, are generally in a better shape than women, may be because they instinctively follow the rule more closely.

A biological predisposition to live long is often suggested. This certainly does exist although the exact reason why is not yet understood. That should surely be discovered, the corresponding trigger has already been found ... but for only the nematode worm, alas! This worm benefited a three times[28] extended life after a genetic treatment. Human would have to wait a bit longer, but a gain of 5 years in the mean life by only better controlling the cancer process, for instance, is a target which could likely be soon reached.

Whatsoever, the importance of the occurrence of genetic effect is evaluated at a level of only 30%. Genes do not control everything but they assuredly contribute. Jeanne Calment, the French famous centenarian who died at more than 122, represents a typical example; her genealogy displayed an impressive succession of vigorous elderly people. However, she has been living stress free with a moral of steel, which has certainly been of some influence for such an exceptional longevity.

Culture and intellectual activities are required, not only to keep a convenient hygiene of life but also to correctly intellectually anticipate the impact of aging. 20% of our population (especially young people) do not know how to read fluently and, still more, write; not to speak of "education". It, then, is normal to find a higher proportion of uncultured and

[28] Last news: the improvement would now be five times!

obese men whose longevity is lower in this population.[29] They also do not know how to acquire the conditions for a harmonious end of life.

The country where the life is longer is Japan, likely because of a better balanced diet which is also better adapted and respected. In the Japanese culture, retirement is perceived as something shameful. In Europe, Sweden holds the record for longer life, whereas the USA lies at the bottom of the pack, their life expectancy stagnates, likely because of, here too, a disastrous diet.

* How to live longer tomorrow?

Without entering the hypotheses of the accelerating technology, how could we grasp the future, on an equal footing?

Statistics says that ineluctably the seniors (so named today to refer to "certain age" persons) are to be more and more numerous ... and more and more valid (we hope so) especially those who "benefited"[30] from an active and balanced life. Currently, in France, nothing has been done to handle the subject; seniors are forgotten and pushed to retirement as soon as possible; many, as a matter of fact, aspire to enjoy a new life of liberty (and idleness).

This social exclusion of the retirees leads to generate a population of idle people very active by their consumerism

---

[29] We shall not discuss here the causes and responsibilities this implies, which would require a longer debate, but we shall limit us to the facts.

[30] There is, here, a social debate which does not enter the framework of this book.

(travels, cruses, retirement homes, and distractions brings up an economic sector very prosperous), at the same time, they could be an unexploited resource of productivity: by not utilizing their skills, there is a loss of the gained experience, loss of a qualified human potential at a time when all energies should be brought together. Such exclusion adds a significant burden to the social welfare budget.

On the other side, Swedish, overall champions of life expectancy, envision retirement only in case of unmanageable physical failure and in any case 10 years after the French. By contrast, they take great care of the "recycling" of the seniors upon reaching 50 years (thus preventing an unproductive unemployment) and of the planning of their living to keep them in their activities or reorient them in another more current perspective. The recycling of the seniors is also a confirmed element for the progression of the longevity and good health. It is with good reasons that they live longer and in a better shape!

* How to do better?

If we take into account the new principle that death is not mandatory, and that we only die from a disease, then it turns out that diseases must be eliminated one after the other as soon as they come or even prevent them by eliminating the causes. Genetics is able to alert us in due time. The aging of the organs itself could be considered as a disease. Nobody knows exactly the details of the chemistry but it is likely an underlying molecular oxidation mechanism which is involved.

The DHEA hormone of Professor Baulieu was ineffective but Professor Shinya Yamanaka (Nobel Prize winner in 2012) does promise a cellular rejuvenation reprogramming which could reverse the aging phenomenon. One already knows how to do with isolated cells; we, soon, will know how to do with constituted organs.

Axel Kahn[31] says that old cells are to be eliminated and replaced by young ones. Miroslav Radman expects to replace or regenerate the "damaged part" when necessary. Everything converges towards a new broad spectrum of possibilities which are to improve our longevity in a near future.

We are unquestionably heading to build a huge centenarian population (men but namely women). Lengthening the life and maintaining the seniors would be the means to raise the age pyramid to prevent it to become too much bellied!

---

[31] Axel Kahn, *Un chercheur en campagne*, Stock, Paris, 2012.

Chapter 7

Towards What Kind of Research?

What could be the future of the current scientific research which is spreading its results in all directions, among all societies? This concern tends to get at the foreground of the human activity; it is likely to be present at all levels and in every sector even if it is often not so "scientific". Would research become an essential activity extended to the whole planet and all social levels? Can it be considered as a goal for civilization?

From the lab to the everyday life, there could be a gap

From the original discovery, the serendipity, and the scientifically validated invention, the implementation of a technical novelty in our daily lives, it takes sometimes a shortwhile under right circumstances but if things are not favorable, it can as well take a long time if it requires a far more profound reappraisal of the infrastructure or the established habits.

Such a facilitated process was observed with the outbreak of a mobile which quickly spreads at a worldwide scale. But it will certainly not be the same for the self-driven car. Indeed, if the Google's experience has proved positive, future projections are quite staggering and could, in any way, arise in the daily life as by the wave of a magic wand!

The master plan[1] is intended to reconsider not only the road car traffic but also the complete scheme of the road transportation in the US: trucks and lorries will be operated by groups or "trains" of several vehicles associated and self-driven in a convoy, 30 cm apart, at a speed of 150 km/h. Drivers being idle all the route along, will be busy with their commercial management works in their cabs which could be transformed in a computerized office. So, in such a situation, why should the driver be necessary if the truck is able to find its way by itself? Holy GPS, pray for us!

Obviously, such a daring project is very attractive owing to the reduced risk of accidents,[2] the substantial fuel savings, and the improved mean speed due to a well regulated circulation. However, the orchestration of the movement of such a vehicle fleet in the global circulation will require a complete restructuring of the total road infrastructure of the country. If such a project is to be run, it will take a long time to be established, it will require massive investments and a full

[1] *Fortune*, November 12, 2012, p. 17.

[2] Software bugs excepted!

reconsideration of the work habits. That is the perfect example of an advanced technical project the feasibility of which is almost acquired but its implementation remains far from granted.

Knowing and be able to, belong to two different kinds of being, not connected at all to our present society. Jumping from one to the other is in no way obvious. From the validation of an idea, the discovery in a lab to the implementation and the extensive use there is a gap not easy to be crossed and this is not intrinsically to be condemned as many discoveries need to be carefully evaluated and tested in their near-by or remote implications before being released and this is to be more and more available. The present example of the "shale gas" is especially illustrative, with all the reckless passions it generates… and also the underlying tremendous interests.

The present society is composed of very diverse humans. Some are educated, ready to reflect on the existential problem, others would have to be informed and taught of the possible evolutions, many will have to wait and submit. Who is to give an advice, who should decide? This diversity is however the basis of our democratic civilization[3] which needs Nobel Prizes as well as a simple worker, each his own role. One cannot imagine a society composed of only Nobel Prize winners; they are rare and that is fine because that is what makes them valuable.

---

[3]The notion of democracy has been largely distorted in respect of its original Greek origin.

The penetration, the democratization of the new technologies could arise through the need, the persuasion, the commercial promotion, the legal obligation or the constraint depending on the circumstances.

The place of the individual in the society will only depend on his "functionality", that is to say, his immediate functional ability to fit with a job profile, without taking into account, as in the past, culture, education, know-how. So will operate the future Directors of Human Resources (DHR); but there is already some truth in that, if we refer to what happened in San Francisco for some time.

In this famous town, at the edge of the Silicon Valley, serious social tensions appeared with the massive arrival of billionaires of the "New Tech Class" and their accompanying highly qualified technicians (highly paid, high competence, employment volatility). Prices rocketed, real estate exploded, thus forcing the ordinary citizen to move towards suburbs and inciting hatred against those who "have managed to succeed".[4] This is incredible in the US! Banners were held which read *"F... k off Google"*. The success sometime becomes volatile as technology or the markets change and the functionality disappears.

Anyway, Silicon Valley (annual *per capita* income: $63288!) is the wealthiest part of the wealthiest state of California. This is to the point that it claims for independence inside California.

---

[4]Disrupted, Katy Steinmetz, *Time*, February 10, 2014. Time to talk about the Inequality World, Rana Forohar, *Time*, February 10, 2014.

However, the relationship with the scientific news sometimes follows strange evolutions. In the end-user computing the interests and the behaviors have drastically changed since a decade or two. At that time, not so far from now, my favorite tabloid each week talked about programing. It even happened that it talked "machine language", which speaks for itself!

The current scenario is very different; techniques disappeared from the news; people are expected to previously know how to use the new tools which are to be friendly. The only problem is to conveniently guide the "geek" in the tangled web for the choice of a tablet or an iPad or a GoPro or a miraculous app among a wealth of offers, knowing that in a short while a new version with improved specification will be available, which will replace the object you just bought.

However, we should become familiar with the changing and abstruse language of the new generations to wisely choose the app, the widgets we need. The Smartphone has become a labyrinthic system, the average citizen (adult) only knows a very small portion of the possibilities.

The research and its evolution

The way research is performed is quite different in the domains of life sciences or that of the so-called "hard" sciences. That could be illustrated in a parable.

In the first case, contributions are cumulative. Each researcher brings its own little (or big) stone which takes its place on the common road. Going backwards is allowed, the stone is still there among others. The biologist keeps his know-how. The behavior or the metabolism of the drosophila becomes more and better known in the details, but the fly is still there, with other mysteries to be clarified; knowledge accumulates.

The second case is different for the physicist. He works in an exploratory manner in a universe he creates all along his way; he has to jump from a stone to the other, bring also his own but there is no time to see where it lies because, when turning around, he can see, on the track, the "asphalt finisher" of the technology which comes and changes the landscape with its tarmac. There is no more stone but only a smooth road; the stone has been "swallowed"!

In spite of this difference, the convergence of the sciences now is evident and the way researchers work get closer, inexorably. The road tarred by the physicist is now open to the biologist who, moreover, already sees, far away, his own "asphalt finisher" with hungry data bases. It is no longer a time for the "butterfly hunter" it comes to the "data crunchers".

Medical research and the computer

In the medical domain, the research has gone in a different and particular direction until some years. The process of the search for remedies to our problems was always the same but things are changing.

The classical approach

The doctor, in front of his patient, inevitably begins, with a glance, his visual appraisal (man/woman, old/young, tall/

small, stout/thin, etc.). He obtains, at this initial level, a subjective opinion enhanced by his past remembrance of the patient and his own medical knowledge. Then he comes to a more detailed and updated questionnaire to further investigate his patient; all of this is purely intuitive, poorly measurable, not quantifiable at all. Instead, he could get help from a medical examination or biological analysis to guide his judgment and lead to a diagnosis.

One can sketch a more or less simplistic case as "the patient cough and sneeze"; this observation is, then, complemented with a short enquiry on the recent or more remote past: one search after anomalous observations which could be related to the present status. Then comes the confirmation of correlations with the disease: "he sneezes because he got a cold yesterday". The most likely cause is identified so that one can look for an adapted remedy.

The entire story follows a scenario of "cause and effect" mechanism which results from similar observations in the past and correlations previously established. The identification of the problem is approximate, intuitive, and statistical; the selection of the remedy is not obvious, it must arise from personal, vague, variable criteria.

This simplistic and sketchy example is indicative of every medical procedure to provide an adapted therapy. At a higher, more systematic, if not scientific level, a laboratory specialized in medical research somewhat processes in the same way. Similar cases are identified, aggregated into classes and the effectiveness of the therapies evaluated from statistics and trials. A global vision can be acquired which helps to determine possible subcategories in each category; thus corresponding to a specific, diversified and identified medication. A classification of the respective therapies is then transmitted to the doctors and updates the previous catalogue in which information is recorded. Sometimes the evidence shows that the conclusions are to be revisited or that new and more adapted drugs or procedures are to be taken into consideration, which is an endless back-and-forth process.

This is how vaccines, antibiotics, or the Koch's bacillus were discovered. Generally speaking, such a procedure does not only rely on a unique case but requires an extensive analysis of multiple observations of similar cases; it is only after having assembled a great deal of converging information that a lesson could be drawn.

One identifies the subjects to classify the symptoms, and note the circumstances. Finally, these observations get accumulated in sheets in a folder from which a dominant trend is to be expected. Statistics are then performed, guided in a more or less intuitive manner.

It turns out that this way of doing, somewhat groping in the dark, leads to accumulation of hundreds or thousands of cards and the way to find common trends in such a diversity, quickly becomes acrobatic, thus exceeding the capacity of a human brain (even computer assisted). Many subtle correlations can escape analysis; many wrong tracks can be generated. However, at this level, the computer might also be very helpful to classify, memorize, sort and retrieve information. In spite of that, it all remains at an inevitably limited scale of a human who keeps the final decision.

The new approach: Big Data

It turns out to be a very different situation when dealing with a "big" computer which is not afraid of "crunching" mountains of data. It only needs to carefully explain what we are trying to do, how to search and the way to display the results. The computer will do its job consciously, quickly, logically, error free. Accordingly with Larry Page, every problem, as complex it could be, must find a solution if it is reduced to a large enough amount of data and if a sufficiently large computing power is projected on these data.

The complexity of a problem, whatever it is, always results from its multi-parametric nature. When a single parameter is to be taken into account its solution is straightforward: the physicist would so explain easily any one-dimensional problem such as, for instance the falling bodies. On the contrary when numerous parameters which interfere together, are to be considered,[5] things become quite more complex. The more likely solution then comes from calculations and statistics. The human brain is overwhelmed and only the big computer is able to get through the enormous mass of the recoded data.

It was the case for many problems already addressed with some success, such as weather forecasting, economy, simulation, neural networks, genomics and even crime and law enforcement! The Big Data phenomenon is likely the biggest software challenge of the decade.[6]

---

[5] Such as the famous "many bodies problem".

[6] Michal Lev-Ram, *Fortune*, September 15, 2011.

There is now a radically new medical approach based on the massive analysis of accumulated data, far beyond the human possibilities. The two leaders of this new philosophy are obviously Google (specialist of the search engines) and IBM (specialist of the big computers). IBM has placed its bets on its parallel structure computer Watson[7] which was adapted to oncology studies and would be able to answer any question in the field, thanks to a "cognitive computing"[8] software which makes it able to understand all what is told to it. Its intellectual power already surpasses that of a human brain, as it was said, and it learns much faster.

The individual medical file is then poised to scale up substantially, integrating the medical history, with its positive and negative elements, the known predispositions of the individual and his family, his genetic patrimony, but also every detail concerning his own personality (diet habits, addictions, way of life, profession, family, etc.); a real police investigation easily documented through the Net. Google knows everything about us!

By now the research programs are directed towards two directions: on the one hand, cancers (beginning with the lung cancer) and on the other hand diseases related to cerebral degeneration (Alzheimer and others). In each case an impressive number of patient's files were gathered (several hundreds

[7]The famous computer Watson was recently shrunk to the size of "four pizza boxes" following IBM.

[8]IBM's massive bet on Watson, Jessi Hempel, *Fortune*, October 7, 2013.

of thousands) in a gigantic data base. With all of that data, the computer, conveniently teached with subtle software, will search for coincidences, tendencies, analogies, and statistics, will look for possible explanations to get a global synthetic picture which could be within the reach of a human brain deduce practical and usable conclusions.

It is already known that cancer is not unique but divided into several different specificities which do not follow the same biological process.

It is already known that proteins[9] with their interactions with the genes do play a role in these affections. But there are some more than 360,000 different species of these molecules in the blood; we need to find out which among these are active, with what combinations, and what kind of genes are affected? The problem has already been tackled in the case of the colon cancer.[10] There are questions the computer will have to answer.

Alzheimer was until now considered by the specialists as a "deep well of ignorance".[11] Two years ago I asked a doctor who is a well-recognized specialist of the Alzheimer disease. He told me: "We are ignorant of everything, the diagnosis is uncertain, we ignore the causes and origins; we know nothing

---

[9] Proteins are organic macromolecules composed of amino-acids.

[10] Could a super computer beat cancer? Brian Dumaine, *Fortune*, September 2, 2013.

[11] http://www.sciencesetavenir.fr/sante/20131224.OBS0611/alzheimer-son-point-de-depart-dans-le-cerveau-localise.html.

about the mechanism; we don't know exactly where it lies and what part of the brain is involved; so there is not and there will never be a possible cure."

However, some light was recently thrown on the triggering mechanism and evolution of the disease. Two protein molecules were identified which could be at the origin of the disease, which make neurons dysfunction in the lateral endorhinal cortex to propagate towards the parietal cortex (center for orientation) and then the hippocampus (memory center); here too, the computer is busy. There is still no treatment available but now we begin to have a better knowledge of the enemy, we know where it hides and how it is acting, that is already a lot and it is just recent.

All of that has become possible with the use of a big computer and importantly "data crunching" in huge databases. Now, difficulties arise.[12] The available archival information comes out of disparate sources (interns, oncologists, surgeons, radiologists, pathology labs, etc.) and under non-standardized forms and supports (sometime written by hand and scanned!) with variable methods of analysis. All of that makes an organized exploitation rather difficult (unstructured format); on top of that doctors and hospitals are rather reluctant[13] to normalize their "Electronic Medical Records" (EMR). The infinite diversity of the types of cancer and corresponding proteins also does not help. The problem shows a tremendous multi-parametric complexity.

[12]Can big Data cure cancer, Miguel Helft, *Fortune*, August 11, 2014.

[13]They use to say: "EMR sucks!"

A young Californian company[14] generously supported by Google[15] nevertheless tackled the problem, creating a "hybrid-human machine learning system" with 50 nurses who manually entered the data of 500 patients. The corrected data are entered into the computer in a standard form in order for the machine to learn and detect the errors by itself! Other companies have attempted the challenge but failed to succeed. An immediate victory is not for certain but the efforts are underway; the delay to succeed could be a decade.

The new dimension of research

Research, generally speaking, is no longer the affair of one man, as genial as he could be, neither of a team of qualified scientists. It is a matter of budget, financing, maturities, collective instruments that are often expensive and go quickly out of date. The university laboratories (in Europe especially) assembled around a visionary boss, does not make any sense today. This does not mean that the original idea is of no relevance, but it has to be justified in advance, shared between all of the contributors; it will be immediately "consumed" if valid, otherwise it will be forgotten. The dehumanization is achieved but the fun of the researcher is still alive.

Obviously, the US still keeps ahead of the research pack. Their Universities have adopted a very efficient model of organization, providing a large and stimulating freedom of action and financing.

---

14 "Flatiron Health" was founded by Zach Weinberg and Nat Turner.

15 Google Ventures into Investment.

The researcher keeps an excellent social recognition, he is well paid, he can realize his full potential, but he also has to succeed if he wants to carry on his work. That is, here too, the harsh law of the natural selection: the loser, if he continues to lose, will be mercilessly rejected.

Reciprocally, the researchers who "find", immediately benefit from every kind of facilities and financing to promote the applications of their ideas. No one will reproach them if they make money or possibly become wealthy, quite the contrary, they will get the respect from fellow citizens for their contribution.

Contrary to what people may think, the American Universities[16] are not so big as for the number of the students, what perhaps makes their quality.

Apart from the Universities there are also many research centers funded or not through federal or state budgets over targeted goals (DARPA, NASA, NIH, military, etc.) but sometimes through private foundations, for instance, the "Janelia Farm"[17] in Virginia.

The recent story of the company D-Wave in California is worthy of mention here because it is typical. D-Wave is a startup created in 1999 by the Canadian Geordie Rose from the utopian idea of the Quantum Computer. Such a computer

---

[16]Stéphane Marchand, La ruée vers l'intelligence, Fayard, 2012.

[17]Foundation created by the billionaire Howard Hughes which is devoted to the studying how the brain treats the information. The foundation provides the researcher with everything he could need. The aim is precise: to entirely reconstitute the brain of a drosophila!

only existed in the mind of some scientists, following a purely theoretic remark from Richard Feynman in the eighties. Nobody had any idea on how to make it neither how to program it; everything was to be invented from scratch. That is to say that this project was without any hope to have it achieved, even in the long term.

All of this did not prevent D-Wave to raise a noticeable capital venture that made it possible to finance the company and its 120 qualified employees until now, without any income. This is the miracle of the American free enterprise. NASA contributed also with some universities and other partners. Today of course Google, which did not exist 15 years ago, came to join the list of sponsors with the enormous financial means we know.

In 2013, D-Wave sent to Google (and NASA) the first quantum computer of the history: "D-Wave-Two" which contains 512 supra-conductive gates on a Niobium tape at a temperature of … 20 milli-Kelvin[18]!!! The price was 10 million bucks!

Theoretically, the machine could be capable of performing an astronomical volume of 2^{512} operations simultaneously, that would put it well above what we could hope to do with a conventional computer structure, even a highly parallelized structure.

Nobody, however, is able to say how the "adiabatic quantum computing" could work, neither if it works because the implementation is really acrobatic. The secret is well kept.

---

[18]The absolute zero (roughly −273°C) is the starting point of the Kelvin scale.

Nobody can say what kind of problems this machine would be able to treat because quantum calculation is performing for only typical issues concerned with gigantic data amounts among which a common solution is to be found.

In brief, detractors abound and applications still are to be discovered, but if it were to work ... that should be quite a revolution!

In France, universities evolve in a less idyllic climate; the researcher with a "genuine caring" attitude is considered poorly by the society, somewhat as an uncontrollable parasite (nobody understands what he could do, neither what he is paid for). At the most, he is tolerated if he makes his hobby low-key, because research cannot be anything else than a costly fantasy. He has to secure his living by giving lessons to mobs of non-selected students. He is not called Professor, as in other countries, but more simply "teacher-researcher" (without upper cases) which is less elitist, more democratic but more confusing.

His professional life is an uphill battle between his search for finances, difficult management of a laboratory submitted to a bureaucratic and hostile[19] administration, various nasty occupations, and petty rivalries against who succeeds. If there is still a bit of time, he may carry out a little personal research project, surrounded by students who are in a hurry to get their diplomas whatever the results could be.

French universities must be egalitarian, where selection should be banned. The position of the researcher is not so dif-

[19]One day, I was rebuked for generating too much work with the industrial contracts I brought to the University (which yet charged its own taxes).

ferent in the framework of CNRS which however gathers substantial and permanent human resources. The financing of the laboratory is mainly obtained from the government, but both public or private organizations may also come forward to finance. However, none of them would take the risk to integrally support the project but an only fraction. This makes the program more difficult and slower to implement, because of the necessity to find several contributors. A final report is often required but not always analyzed and rarely expertized.

The relationship of the researcher with the industrial world is not so easy. The person from the university is considered as a by-product, and therefore less respectable, in comparison with the selected graduates from the "Grandes Ecoles"[20] which do not have the vocation to produce researchers but more simply engineers; these two kinds of jobs are not so similar: the engineer has to tackle immediate questions whereas the researcher is to ask and answer questions to be raised in future.

Another difficulty to be overcome (and this one is dramatic) comes from the complete absence, in France, of a patent culture, even in the industries[21] people consider contemptuously

---

[20] Grandes Ecoles are purely a French invention, not well known in the international communities because their diplomas do not enter the international scale.

[21] One may remember, for instance, our plane "Caravelle" which was a remarkable innovation with the two reactors located in the tail; but Sud Aviation, at the time, neglected to patent the idea.

practical applications: the researcher has to remain a pure soul, hashing over ideas which he alone is able to follow!

About 30 years ago, a researcher who tried to develop a practical application of his ideas was considered a "rogue" whereas the one who was stubborn to the point of creating a "start-up"[22], was considered immoral, financially motivated, unworthy to belong to the intellectual university world. Hopefully, things are to change now.

French Universities are still unable to display a coherent politics of patent management is spite of the assured potential wealth of their source of scientific expertise. Obviously, with this in mind, the exchanges between University and Industry often stop at the level of empty promises.

The relationship with the society in general is not easier. The researcher does not take readily his place in the social life. One may remember Professor Luc Montagner (Nobel Prize winner and discoverer of the Sida virus) who was retired automatically from his post by the University Administration which estimated he was a bit too old. But he was immediately hired by the University of Jiao Tong in Shanghai where he was allowed to pursue his work on the biological nanotechnologies! He was also offered a dedicated institute which bears his name!

In spite of all that has been said above of the difficulties and the sad picture that could be drawn out, Science, in France, is nevertheless wealthy and involves an increasing number of

[22]I do have made the experiment with some other "adventurers".

highly qualified persons; it valuably contributes to the world-wide efforts in every branch of science and at the top level.

Mathematics especially constitutes a more individual science and that is very convenient for a French educated mind; it is no wonder that, in this particular domain, French researchers are bright and are often granted Fields Medal which is the equivalent of a Nobel Prize. Such a distinction, alas, rarely hits the headlines of the French newspapers which may be keener to advertise a soccer player!

By now, Science irrepressibly spreads over the world and no longer is the purview of our traditional occidental countries. It begins to concern many countries, far away, contributing to or reinforcing local politics. A surprising example came from Qatar, which is sinking under a mountain of petro-dollars, which bought the collaboration of most brilliant researchers (essentially Americans) to create on its soil a scientific Arab center of excellence, thus renewing with the antique Caliph's tradition and always with the same and only argument: money. The temptation is strong for all these scientists to get a well rewarding job and at the same time unexpected facilities and working tools.

Today, alongside the traditional countries (Europe, USA, Japan, Israel, and Korea) new challengers appear from the BRIC (Brazil, Russia, India, China) but also from smaller or emerging countries we were not expecting: Turkey, Iran, Tunisia, Singapore etc... Scientific research is now a world concern; it is intended to be the promise of prosperity, life quality, globalization integration; but this entails the major drawback of a broader knowledge-sharing which could be dangerous if put in irresponsible hands.

We are here to evoke a particular case which is worth a special development: China.

This country of a millenary civilization, a long time closed in upon itself (Middle Kingdom), is incontestably at the very origin of all of our civilizations (except original Americans). Every important discoveries (paper, compass, printing, bank-notes, rudder, black powder) do belong to a Chinese origin, even if others unduly claimed for the paternity; but Chinese never exploited them.

China has now awakened[23] and promises to become an indispensable partner in every sector including in the domain of scientific research.

Yet, the Chinese culture, the organization of thought, the intellectual disposition arising from the language and the scripture (ideograms) give to a Chinese mind an approach rather different from ours which we acquired from the rational traditional Greek culture: Chinese language is more descriptive and less deductive; it works more like a camera than a computer (if I can dare this kind of metaphor!).

After having been, for years, the "world's workshop", China has been enriched with its foreign oriented trade and has now the means to reach higher stakes.

---

[23] "Lorsque la Chine s'éveillera... le monde tremblera", said Napoléon. This citation was taken up by Alain Peyrefitte for the title of his famous book (Fayard Ed., 1973).

There is a huge and highly qualified[24] human tank which is able to produce a highly qualified scientific population, as long as a developed university structure is to be built and this is what is to happen in China. Many new universities were recently created; they still suffer from lack of experience and history; but the numerous Chinese who migrated to Europe or more likely the US during the disaster of the Cultural Revolution to conduct their studies are now coming back home; they often became first rank researchers and bring with them their occidental culture; they are called there "sea turtles"! They are often accompanied by some American or European defectors (as Prof. Montagner) to whom the Chinese government offers "golden bridges".

All of this bodes well for a future to come where China will hold its place in scientific innovations; it can also be added that China, different to France, develops an unbridled patent politic which at the moment looks somewhat disorderly, but the will is clear to make it a tool for progress. No doubt that, in China, experiments will be completed without taboos. Some say that the level of research in a country can be intuited from the number of patents and Nobel Prizes, China is preparing for that conscientiously.

While the "ancient world" stays locked in its contradictions, the new Asian world has got a philosophy made of

[24] I was allowed, many years ago, to give conferences in a Chinese university (X'ian) and I was largely surprised by the mental alertness (and impertinence) of the young students who never hesitated to ask justified questions directly to the lecturer thus showing a large maturity. Quite a different attitude in respect of Japanese students.

work, production, enrichment. Intelligence will likely follow the economic expansion.

Half a century ago, Science was fundamental and individualistic centered on the mathematical modelization of atomic, sub-atomic or photonic phenomena. Today Science has become collective, multi-disciplinary, and pluri-national, with a strong orientation towards immediate applications. The rules of the game have changed and the number of implied researchers is much larger.

Nevertheless, Science should remain an individual pleasure… it is not to be easily shared with non-scientific people.

Intelligence and the computer

The Man, by himself, and with the computer, has not (yet) become more intelligent or more educated; the respective proportions still are the same in their respective levels. Of course, he became more qualified in the use of more complex machines but the software "assistance" goes on everywhere: the car's wiper automatically starts when it rains, the photocopier not only photocopies, as required, but also digitizes, identifies, sorts and classifies the documents by itself; we pay with a card, until we got a chip under the ear to check out without stopping!

There is no need now to "know" the object we use, as complex it could be. Moreover, the user's guide and the notices completely disappear; we are supposed to intuitively be able to use the machine we just bought, without the need to learn. Nobody is to ignore but nobody is forced to learn.

Our clothing is solid, crease resistant, cheap, hypo-allergenic, some say they are even to become "intelligent".

The Man himself has changed; he has physically flourished (with the untruth counterpart of obesity); records and sport performances got unprecedented heights, up to now, in the extreme sports and women do have their share. The notion of learning and craftsmanship has been largely replaced by the notion of "pedagogical progression"; there is no longer a question if we could be able to drive a sport car, or speak English, we just ask how much it costs, the result being obvious (in principle). We even find normal that a blind person could fly an aircraft! Yes it is possible!

Internet is there to open the doors for every possibility, but "Internet for all" is just 10 years old; what will become of it within next decade? This is a provisional episode, just as the first combine harvester or the first fridge. Upheavals are there: scripture and drawing are giving way to digital keyboards or tablets and touch screens.[25]

An important element of the evolution of robotics relies on the natural behavior of our children concerning the robot; they "accept" it and use them to play around (toys); quite a different attitude from that of the adults who are more reserved. There is no barrier or restriction. We often are afraid of the behavior of present youngsters but we have to think regarding the coming generation. It cannot be found worse

---

25 — Nevertheless, some are reinventing the hieroglyphics with the pictograms and more recently the Emoji (絵文字), and consider it as a possible language for the future!

than the "hippy" generation of the 68 which preached a systematic challenge and "sea, sex, and sun"!

The day when robotic intelligence will work, and it will one day or the other soon, it will crush the human intelligence without difficulty or remorse. The main advantage of the robot is that it does not feel any sentiment, never forgets anything, nothing is lost, there is no need to repeat again; it is also able to transfer his knowledge to others. Men, even the more universal minds, always have a memory and an intelligence which is limited to the subjects they have been educated in, each one in his specialty. Internet, for its part, knows everything about everything (provided we know how to search) and Google is there to help, crunch the accumulated data and get its science.

Many very complex and sophisticated AI software of the last 10 years have evolved from an experimental status of pure research to the one of practical applications for the general public. This is particularly true for virtual image handling and computer games.

The intelligent machine (if it could be) has its own advantages:

— Its size, contrary to the human brain, has no limitation (A theory[26] of the old times tried to relate the human intelligence to the size[27] and shape of the brain).
— Its higher communication rate.

---

[26]The Paul Broca's theory of phrenology in the 18th century supported the concerns of Balzac.

[27]The size of the brain being limited, there is an essential compensation mechanism which is intended to save the neurons: oblivion.

— Its adaptability to the intended task. Several machines belonging to different specializations can be assembled in parallelized or serial clusters in order to enhance the efficiency.
— The easiness of duplication to get a population of similar or complimentary machines with shared knowledge.
— The unbeatable rationality to optimize a task.
— The availability and the low running costs which are limited to the electrical consumption.

Last but not least, the machine is never tired, it does not claim for anything, it never joins a union, it is able to work night and day without any coffee-break, It does not know (unless taught) weariness, fear or moods and most importantly, the significance of money.

One generally considers two distinct kinds of AI: general or narrow intelligence. This last category is characteristic of the robots (this makes the difference also with an automation or with the human brain). This kind of intelligence does not proceed from anthropomorphism; its own logic is sufficient to activate expert systems in all of their versatility. In its "forecasting" version, it is very good to appreciate the weather to come or the stock exchange; there have been continuous progresses in such domains and they easily surpass the possibilities of the human intelligence.

The general intelligence is closer to an imitation of the brain (WBE)[28] and thrives in external contacts. This kind of

[28] Whole Brain Emulation project.

AI develops a power of optimization as a function of our aims. This can be extended to the possibility for the machine to reprogram by itself depending of its learning.

In this way, Gödel's machine of Jurgen Schmiddhuber (2007–2009) would try to become optimal; the project deadline is not immediate nor in the mid-term, but the day the goal will be reached, there should be a serious problem for all the jobs related to intellectual work today reserved for humans.

This will be still more serious if we reach the ultimate goal of the AI, I mean a "super intelligence" machine which will provide an achieved dominance in domains like mathematics (Polymath Project run by Google), engineering and sciences in general. In such a circumstance there will not be as many possible bargains with the machine as it could be with a monkey, the specialists said. A "friendly intelligence" would then be to invent which would limit itself to the defined goals and provide us with a more pleasant life as the Swedish philosopher Nick Bostrom[29] suggested.

An intriguing problem still remains. Do we know how far such a super intelligent machine would be able to take us? Would we, then, resort to the ultimate argument as Dave[30] did in "Space Odysseus", with its computer Hal: unplug the machine, If this is still possible! This is quite an existential

---

[29]Nick Bostrom is at the origin of the World Transhumanist Association. See: "Are you living in a computer simulation?", *Philosophical Quarterly*, Vol. 53, No. 211 (2003).

[30]Stanley Kubrik's film "Space Odysseus" (2001).

threat because the machine would likely not give us a second chance.

Maybe, meanwhile, also, that our brains would be directly connected to the machine which so would be part of us. Who could be who? Who will command who? This is a diabolic spiral!

However, our coupling with the machine is already on the way. It will be soon be possible to replace our obsolete eyes with a more versatile camera. The first step has been achieved. We would then be in a single level with the virtual world of the machine. Google proposes glasses which offer this possibility without the necessity to replace the eyes (hopefully!).

The preliminary attempts to make a machine understand our thoughts and transfer them to another machine are already fulfilled. One now knows (but everything do accelerate) how to order a robot by decrypting the electromagnetic waves detected from the brain and also by the imaging of the activated zones. No doubt that progress could happen in a near future in the restitution of the "cognitive map" of the brain. This will vigorously quickly get through the reverse engineering of the operating mode of the brain, that is to say the understanding and modelization of the corresponding mechanisms.

The weakened theory of evolution

In its immediate intentions relative to the management of living species, the natural selection revealed by Darwin intends to improve the genetic heritage of the species and its environment adjustment by discounting every "defective item" randomly generated. The evolution only proceeds from the destruction of what does exist but reveals itself to be unsuitable.

That is the way it was of the human species until man successfully began providing care, hygiene, dietetic to the failing babies. This transgression from the natural laws is inexorably to lead to a not yet eliminated accumulation of the genetic deteriorations generated by the nature in a random way. This deterioration should be allowed to continue, but is not yet very perceivable because it spreads over a long period of time; but its acceleration, dragged by the recent progresses, will never cease to amplify thus requiring in the mid-term to take "remedial efforts".

The improved knowledge of the genome and the possibilities now to change it would soon allow such an intervention. Serious questions would then be raised or, may be, the necessity would appear for moral transgressions which could bring up tragic memories. All of that would be replaced in the framework of a predictive medicine, accessible to all, if possible

The first possible intervention could be a prenuptial[31] control of the compatibility of the spouses and the implementation of a preventive abortion in the case of an anomaly. This is already in use for the 21 Trisomy. The second possibility

---

[31] This supposes, of course, the sustainability of the marriage institution also with that of the natural reproduction methods.

would be a direct correction of the genome to compensate the detected deviations; this will constitute in a deliberate manner to replace the old Darwinian equilibrium by a programmed selection, an ideal improved model. One can easily imagine the passions that could trigger. Is the man, in his diversity, improvable?

On the other hand, universal solutions would be inevitably very slow to extend over seven (or more) billions of humans; a transitory regime will be necessary where "improved species" will have to live side by side with the "traditional" ones. Inequalities will be patent. What to do with the "improved" (at the least, the firsts)? Will we have to put them in a "Zoo" or a "Fair"? Would we have "improved street sweepers"? Assuming that the "improved man" get intelligent solutions to every problem, could he be able to implement it in practice or will he be powerless given the foretold inaction of the others who will not have a computer in the brain? Intelligent guys still exist; but, who cares? Would the human be changed for good or for a catastrophic fortune?

As stated, these problems arise from a direct extrapolation of the present situation. Would they escape[32] if the "transhumans" are to switch to the plain artificial procreation using synthetic elements? But there we will be in another universe and we should have to "swallow" so many transgressions before! We should return later to the point of the possibility of an artificial "zero defect baby" for everybody.

---

[32] "Faust" already said this sentence: "the old way to procreate, we state it is vain joke."

Among the many possible combinations of genes in a DNA molecule, it was admitted up to now that only few of them play effective roles, whereas the others being of no use at all, This opinion no longer holds; it is now recognized that things are well more complex. Sub lying influences of these "unnecessary" genes were recently detected, more especially as cancer is concerned, which leads to tremendously open investigations. The understanding of our individual "programming" is still in the infancy.

Chapter 8

Could Man Change or Disappear?

The tools we currently use in biology, genetics, as well as robotics or software are to allow a direct access to man's intimacy to the point that his very nature could be called into question.

Initially everything obviously begins with good intentions; this is all about curing or protecting from well-known plagues, but rapidly one comes to forecast and prevent potential ills from appearing one time or another; then, why not make a step forward further; why not try improving our performances in one domain or another; we are now in small stages of increasing sophistication of our knowledge and technical means, in the process of redoing the Man. Where do we stop and why to stop?

The 2.0 generation, on its "cloud", will accept every transgression as it has already moved away from the traditional values of the Judeo-Christian world which, up to now, supported our civilization.

Man, men and the society

Man, as the ant, is a social animal who established (variable) rules for a community life. During the millenaries, he developed what we called Science albeit unequally. Some societies were deeply involved in this move towards knowledge, others much less, not to say not at all. From here, a large disparity of relationship with Science is to be observed between the different communities and even within them.

In spite of all educational and communication endeavors the access to the scientific world is very unequally shared. Whatsoever, we have now entered a severe mutation of globalization; sharing the means as well as the researches will still amplify at a universal scale. Everybody is expected to be submitted to the implications of the "progress".

It is therefore necessary to avoid a global philosophy, idealized images, and shortened judgments. Everybody will not yet be concerned by the technical mutations. One could compare the situation with the lunar conquest: one knows how to get there, one knows it is possible to get there, some privileged few already got there, but this cannot be the reason that everyone can get there for holidays.

It could be the same with the "trans-humanity": some there will open the way; there will be successes, disappointments or failures too; others will continue the journey and have a rough experimental exceptional ride. Although not everybody will become a Superman in the blink of an eye.

We now enter the first phase of the mutation: the therapeutic phase. There is no doubt that one cannot refuse the expectation of curing serious illnesses, even though it could be at the expense of "radical" treatments. We are even looking forward to have access to such medications.

The second phase will be a direct prolongation of the first one. There will appear less pressing extensions of the new means which have just been developed. These extensions will begin to constitute clear transgressions in our way of life with the development of new habits. These preventive and non-essential actions might be more difficult to admit but the movement will be unavoidable. There will be a collective pressure, a necessity.

As a result the preventive medicine and corresponding care will certainly face reluctances but will finally be accepted, as hygiene had to be accepted at the end of the 19th century. This phase will help us indulge the changes and prepare us for the last phase which will certainly be more radical and drastic. This will be the inescapable phase of participation, not to say "hybridization" with the machines.

Reconditioning our societies is not yet on the agenda but it will become in the mid-century. An important difficulty would be to make it accepted by everybody; given the disparate cultural levels!

Not everybody would adapt, there will certainly be people left behind. Would we abandon them by the way side and keep moving forward? How do we manage these huge populations with increasing needs; will they have to be sacrificed in favor of a select few self-made elite? Would a social segregation be established on the criterion of profession, money, competence as observed in San Francisco now?

May be it is there, a rather pessimistic and regressive vision. Is it worth considering the "voracity" of the "public" (all the public) adapting to electronic gadgets (phones, tablets, game console, etc.) that can be bought at a reasonable price but updated shortly after. From all accounts, we will be doing it soon with gadgets that aim to be therapeutic[1] and will be integrated in our bodies.

Beyond the new therapies which are intended to protect against serious diseases, what kind of "improvements" or "refinements" could the human machine expect?

The first idea that comes to mind is that the machine could bring an easier, more intimate access to a well of knowledge, that is to say huge data bases (i.e. Internet or a more performing equivalent). This approach is already well underway with the "mobile" which is able to provide us with detailed and useful information in a click: urban localization (monuments, access map, transportation, environment, cultural programs, schedules and so on). Still more performances are expected with the vocal assistance which makes it friendlier: every object will become "smart": Cars, household appliances, glasses, TV sets, even tooth brush will obey our voice (and vice versa). They are expected to respond in a clear language and give relevant answers and information we could ask for. (I guess; even my wife, who certainly likes speaking, will be quickly fed up with them).

---

[1] If it is fashionable and if it does not require special efforts.

Obviously Google is ahead of the move for the "Internet of Things" but it is essentially a matter of big data's early days. According to Gartner there will be 26 billion connected devices or household objects[2] (not including PC or Phones) in the world by 2020. Nevertheless such initiatives are also actively developed in France which are called "objects communicants".[3] However, a global solution cannot to be reached soon because the problem is unspeakably complex, needing to adapt to the tremendous diversity of situations in the daily life.

Although the technicality gets more and more demanding and the robot aiding gets more and more performing, how can we effectively get rid of man in basic activities? Who should make the robots? Well, other robots may be?

Our society, in the present configuration, needs a large diversity of competences and skills to cope with. Would this situation be changed? Would it be possible, on a discretionary basis, to transfer learning, knowledge, and a memory, to a hybrid brain?

At the moment, men display a tremendous diversity in their nature, their character, their behavior, their sensitivity; one cannot imagine a unique, standardized model. This irresponsibility, this irrationality, this creativity, this fantasy too

---

[2]Household Objects, Matt Vella, *Time*, July 7, 2014.

[3]Rabit Nabaztag and its successor Karotz are proposed by Rafi Haladjian to contribute to the home life. They detect noises; they can tweet, read mails, and monitor the "know-it-all fridge" and so on.

could never be matched in a machine which has no self-determination, or original initiative outside its programming.

Man repaired, augmented, transformed, trans-human or outright post-human were some of the qualifiers we found in the previsionists literature. Would we be pulled forward in this destiny by increasing requirements for competitiveness, budget contingencies, and race for survival? What to do? For what kind of life? The progress has to be paid in one form or another; would we have the means to carry it out?

Would we get to the point where our behavior of "work robots" would be closer to an accepted slavery? Such a tendency is already noticeable today with ever increasing requirements of a professional life which crushes the personal life. The sight of the Parisian subway at 8 in the morning is very evocative in this respect.

Would we reach a point where the world will only constitute researchers and technicians? We nevertheless require people, for a long time, to produce, care, instruct, manage and even provide distractions, and it still may be some time before a robot could do everything, even if their capacity is to be improved and generalized very quickly.

Could genetics allow itself anything?

Genetics is a young science but oh-how promising! Our genome bears all our hereditary message and all our future potentiality. We would incline to attribute it a definitive determinism, an undisputable fatality of our destinies.

Religions, to varying degrees, pull us to the belief of a destiny "written" by a Divine Will. Ancient Greeks could only conceive an oedipal fate enshrined in our very existence. Muslims made certain that only the Allah is to guide us and that every other form of medicine is useless and blasphemous. The same philosophy also holds, true to a certain extent, amongst Jews, whereas Christians take for granted that God is to guide us. Then would the remaining part of our liberty to guide our lives by ourselves be reduced to a bare minimum?

Geneticians are bold enough but in the same time very careful in the conclusion. If it is true that the gene is determinant for the coding of proteins, the analysis of the cause and effect relationship between the gene distributions on the DNA scale and the physical realty induced in our existences is not yet achieved, far from it; the problem is especially complex. Promises are immense but we know that the genome does not really tell the whole story and that considerations about the living history, the environment and so on, have to be taken into account. Billions of different DNA exist through humanity; to draw a conclusion (which can only be statistical) from the sequencing would certainly require time and powerful machines we do not yet have. Correlations are infinite. So we therefore need to keep modest, although clear and bold. Epigenetics is still a very young science.

Such modesty still does not prevent us in some specific situation of investigation, to already distinguish underlying trends and possible extrapolations. Between the scientific unfounded utopia and the definitive skepticism there is a broad range of possibilities for an action which appears clearer and clearer, but it is evident that the shortcut "one gene = one disease" is to be definitely discarded.

Genetic analysis gives valuable indications of predispositions to diseases, troubles or disorders but predisposition never meant an immediate danger of death. A "predictive" medicine is currently developing, starting with the more terrible illnesses and the least mastered: diabetes (type 1), Alzheimer, atherosclerosis, etc... It is no doubt that all this knowledge is to make rapid progress and shall give results. Therapies issued from the genome have already been experimentally checked on humans and these attempts are being developed in precision and security because collateral effects of these treatments are still to be evaluated. It could neither be excluded that new diseases could appear, requiring adequate treatments. The maintenance of life is likely to resume in a permanent fight against death.

In spite of all this careful language, it still remains that genetics will be a decisive factor in the rupture or the discontinuity the GNR promises us and which is already there. Political decision-makers seem unable to take any initiative in front of this expected disruption. It was already the same twenty years ago with the beginning Web.

Some scientists, however, extrapolate these innovations towards a world where everything could be regulated and conditioned by an "optimized" genome which could guarantee an extended life, immunity to diseases, extended natural "gifts", etc. They are "Trans-humanists"[4] or even "Post-humanists" which announce the emergence of a new humanity after crossing a barrier called "singularity".

As a counterpoint here is also "bio-conservators"[5] activists who violently refuse continuing researches in these advanced domains they deem it dangerous.

The well-known trap of the "Science-alibi" is set for us, which justify anything through a scientific varnish. It is obvious that all the intent we are going to encounter would certainly not be benign, whether ideological, commercial or political. We must remain vigilant because the promise of an improved future could lead us to accept the unacceptable.

Politicians, at the moment, are overwhelmed and unable to grasp the problems which are yet to come. There is neither study nor long-term strategy about the subject. Debates are not organized to evaluate the risks and the benefits; the arguments remain vague. Nevertheless it would be foolish to let the sorcerer's apprentices do what they want; they will assuredly be encouraged by the pressure of various lobbies. The globalization of the relationships mean that any regulation (as long as this could be conceivable) would only be at a global scale.

The future that we could expect is full of attractive promises but also serious threats. Simply opening the individual

---

[4]Ray Kurzweil, The singularity is near, Penguin Books, 2006.

[5]Françis Fukuyama, Our posthuman future: Consequences of the biotechnology revolution, Farrar Straus & Girous, 2002.

genomic data to the public might lead to discriminating and perverse use if the access is not limited and controlled by international laws and regulations: An insurer could adjust his fares or refuse a contract following the genetically detected risks of a client; a company could refuse hiring somebody for the same reason or even require a genetic treatment adapted to its productivity criteria.

Where then would the role of the doctor fit amongst his duties such as the concern to heal, the prevention of a disease and bringing a comfort which necessarily is in no way essential? How are we to organize his filter role and ethic? Where is the limit of the common sense, where does abuse begin? Would anything be at the free disposal of everybody or would access be regulated? What right would we have to intervene on individuals who are recognized as naturally dangerous[6] (this already exists)? Would life be possible if we decide, following a precautionary principle to make everybody undergo a medical treatment for all the predispositions (even non urgent) that could have been detected?

One can see here the huge diversity and multiplicity of the questions we will have to answer before giving free access to these new means. A "geno-management" is to be invented to prevent us against the "geno-Knock"[7] doctors who are to appear, as suggested by Laurent Alexandre;

---

[6]Such a question is not new and still has received no satisfying juridical answer.

[7]From the famous old film "Doctor Knock" (1951).

could we refuse to anybody the possibility of enhancing his possibilities?

It is even possible to imagine a terrific possible situation, in a dangerous totalitarian state (there still are examples), where the genomic corrections would become mandatory to obtain obeying subjects or slaves (possibly consenting) programmed to perform a particular task. Would the restrictions which could be operated in democratic countries, be circumvented in the more permissive "off shore paradises" as it is already observed?

In our present society reactions begin to take place, in a mess. The gene has appeared in our scientist societies as a means to protect oneself. Blocking everything is unconceivable, accepting everything is unthinkable. Doctors are very reluctant and fear the novelties; opinions are divided: religious groups (Catholics especially) see an Apocalypse there, extreme leftists show a medieval obscurantism. But all of that does not prevent "high tech" capitalistic Chinese companies to invest; they organize trans-humanist visions for a mid-term future. Our youngsters will certainly be prepared for this kind of life and challenge; their behavior today is prepared to such a mutation into the unknown. The virtual reality, largely developed in the video-games and even projected on the family TV has prepared them not to be afraid of any innovation.

Among the benefits expected from genetics one can obviously find the fight against aging: against the diseases which make us aging in an accelerated process (cellular damages), but also against the very principle of this wear. Promising results were already obtained by geno-therapy on the nematode worm and the mouse, whereas an extension to the humans could be reasonably expected; the main difficulty in the human case remains in the rather long human life, much longer than for the worm or even the mouse; we will have to wait a much longer time to be sure if the medication actually works. If somebody were to live for 200 years, as promised, we should wait at least a century[8] (or two) to be sure! The day when such a means would be available death will be considered unacceptable. Today, already, we are protesting when every effort was not performed.

At this level genes are not easy to control, their effects are indirect, diverse, multiple and they even change with time. The objective to reach, of course, is not only to extend the lifetime but also to do it with the best acceptable (if not improved) physical and mental conditions.

Genetic modifications have just begun to become feasible; we do not yet have the experience of the so-opened possibilities. The will to "correct the defects" of what nature has brought to us will be more and more pressing. Let us suppose that such corrections will be effective and accumulative; could we imagine achieving, in a step by step fashion, "standard models" of ideal artificial genomes guaranteed against diseases, defect proof, and specialized in one or other specialty, on request. Will we then reach a unique family of ideal

---

[8]Presently, the anti-aging remedies are only considered for kids.

DNAs? Would we have to limit humanity to classified stereotypes? One would like to echo the fable: "boredom, one day, was born from uniformity!"

Another possible drift from the eventual control of the genome associated to biology may also lead us to imagine possible perverse deviations, such as terrorism or mafia.

And finally there is also a domain where genetics is to play an important role, I mean reproduction with the multiple opportunities which already have arisen and will certainly interfere in our daily lives. This subject will be more extensively addressed later because it requires a detailed development.

From the repaired man to the trans-human

Will the nature of man have to change or will a new one be fabricated with the creation of an artificial life out of an ultra-powerful technology?

Techno-medicine and genome "surgery" are forecasted to the year 2020, which is not so far from now. The sequencing certainly would become trivialized, with a paltry cost. On the other hand, cellular biology is also progressing thus allowing the reconstruction of aged tissues, the generation of new ones, and the creation of artificial cells for cellular reproduction. And, to complete the picture, computerized implants are emerging.

Trans-humanists are convinced that such possibilities of acting onto the human machine will become generalized; they even see that as a perpetual "updating", similar to that of a software, would be required; they do not imagine to put a limit on the improvements of our cognitive, memory or even physical faculties. With the GNR, they say, the human could be modified and "customized" at will. But we still have to define what would be the need and the necessity. Anyway, the need never is necessarily required: people do exist today who absorb anabolic steroids with an intent to grow their pectoral muscles! Narcissism has no limit.

It is obvious that the first "adjustments" should be purely therapeutic or prosthetic and the only option should be to accept them, they will even be welcomed. Diseases, deficiencies, disabilities will one after the other find efficient and workable solutions. But the acquired technical facilities will also create a desire to go further in expectation of comfort, interest or even simple curiosity. Then, man will be repaired; they may be improved or augmented say the optimists; but augmented to what?

Obtaining a larger memory, more mental alertness, more strength or skill is assuredly laudable; this "doping" could even become essential in the mad rush to the professional competition as it already is to sport. The phase of repairing will be accepted without difficulty even if it could give rise to some incoherent reactions: cochlear prostheses implantations were difficult to be admitted whereas brain implants were accepted without any problem.

Disturbing "tests" have already been attempted: man/monkey chimeras were produced from human embryos stem cells injected in the brain of a macaque fetus. The experiment worked perfectly well giving hybrid beings with mixed brains; but were destroyed before birth.

One would then enter into a cycle of "hybrid machines" where humans become less important given the function he has to fulfill. Man/machine hybridization shall not be long to develop, as the know-how software evolution, the inert/living connection and the AI progress. To what extent can the role of the human be reduced? Would man still become necessary if machines can perform better, quicker, and at a lower cost[9]?

What kind of a man would then be required, what rules to be obeyed, what limits, what deadlines? We shall have to answer all of that. The transformation could be irreversible and could require changing many of our present conceptions of life. Would the notions of happiness or fun be abandoned in favor of superior efficiency?

If we put aside the specific interventions on the genome or the stem cells we talked about before, we would still have to examine the current perspectives on grafts or transplantations. We now enter the Frankenstein domain,[10] which was only a romantic fiction once, but still left traces in our minds.

Grafts are to be distinguished from transplantations in the sense that they do not require vascular anastomosis, that is to say the reconstruction of the blood net. They however

---

[9] In the world of the machines money is meaningless.

[10] Mary Shelley and Susan J Wolfson, Frankenstein, the modern Prometheus, NY, Pearson Longman, 2007.

conform to the same goal: replace a failed or torn off element by another one of the same (auto-graft) or of different origin, animal[11] or even artificial origin.

An ideal solution to these problems of course should be the "salamander solution" (or "lizard tail") which would consist in making the organ regrown by itself as a simple cutting. Things are on the way but it is still too early to imagine a possible human application.

The main obstacle to the transplantation being to maintain immunity compatibility, high hopes are placed on the fabrication of organs grown in an *in vitro* process, from stem cells of the patient. This begins to be routinely done for skin grafts[12] on major burns, for retina[13] or larynx grafts, for the articular cartilage[14] of the knee and is expected to be extended to internal organs like kidneys or even heart.

Current advances are spectacular and no doubt, should go through very quickly leading to even more impressive results which certainly will fall soon into banality.

Let us also think about biological three-dimensional (3D) "printing" for a moment. We are for a long time (some thirty years or so!) accustomed to print our documents, even photos, with printers, laser or ink jet; and it is still taken for

---

[11] Dentists usually use biological materials to reconstitute a bone or a gingiva. Heart surgeons use pig valves to repair failing hearts.

[12] Skin is an organ, in the full sense of the term.

[13] Retina, too, is an organ.

[14] Such "synthetic" cartilages now can be produced by cell growth *in vitro* and then 3D printing.

granted. However these high precision, tireless machines are pure marvels of mechanical, electrical, software, and even chemical technologies.

They are such trivial objects that nobody paid a particular attention to them except when the ink cartridge is empty! However some have developed the technology of this instrument to pilot a string of melted plastic in such a way that the material is deposited in thin, regular and solidified layers based on a software program. This allows layer by layer reconstruction of 3D solid objects, the shape of which are often very sophisticated. These machines, as well, fell into banality; one can find them in public use in some shops, super markets or even post offices; to scan an object in 3D using a laser, a specialized software is only required to get the measurements, the machine takes care of the rest with a reasonable delay.

But this was not enough for the limitless imaginations of men. An incredible idea was born to use such machinery with living materials. A machine has been built which is able to grasp stem cells individually one after the other and place them in the right order following a pre-established program; the attachment of the cells is spontaneous. The experiment shows that the cells easily survive the operation, and then gradually constitute a complete living tissue matching the shape of which has been designed.

Such constructions have already been achieved for perfectly functional cartilages with the exact dimensions and shapes of the natural ones which have been previously measured with a laser scanner. I guess orthopedist surgeons should be very pleased with that! In the wake of such inventions, other tissues have also been experimented, such as liver, which of course are compatible with the recipient.[15]

This is, at the moment, an experiment only operating to make the tests of drugs or other procedures easier in the laboratory. The realization of a complete, functional, and complex organ will arrive in due course of time. The major difficulty currently is to keep active the vascularization of these tissues and their alimentation in oxygen during their growth.

Some have even made plans to use this method to recreate complete organs which could be used for auto-immune grafts, thus avoiding any rejection phenomenon. And if possible, create more complex constructions such as a nose, a liver, a heart ...a living Meccano! And in the process, why not also integrate, an electronic component to boost the functioning of the organ and give it enhanced performance? Alas, we have not yet got to the point where we could hope to create a brain![16]

I just dropped these premonitory lines that I discovered in the illustration of the French press[17]; it therefore seems that the French labs are also on the track of development.

In a more fantasy mode, I also discovered that a Dutch artist succeeded in recreating a 3D copy of the lost ear of Van

---

[15]Such a graft also has the non-negligible advantage that the used cells are new ones; a new, compatible young organ then has been fabricated.
[16]Laurent Alexandre, private communication (March 2014).
[17]Comments from L Alexandre, by Anne Cagan, 01Net, 6/19 mars 2014.

Gogh from cartilage cells grown from original cells belonging to a descendant of the famous painter!

From a surgical point of view, one currently knows how to restore members which have been accidentally sectioned, the more delicate difficulty lies in reestablishing the nervous continuity. So the grafts of liver, kidneys and even heart are becoming routine. More complex operations have also allowed reconstituting complete faces,[18] heart/lung sets or more complex assemblies. One even just succeeded in reconstituting the hip and basin of a cancerous woman. Now she can walk!

Promises of robotics in the domain of surgery and microsurgery are also impressive and have opened up new and bigger perspectives with the development of associated AI. Both the precision and the reproducibility of the move are not to be compared with the human action, which are much more uncertain.

Step by step The Frankenstein's fiction is becoming a reality. This reconstructive surgery could also be extended to improving care of disabled patients and to plastic surgery which is now commonplace, thus, more often than not, bringing a psychological non-negligible comfort.

How far can we carry on the dream or the delirium? Could we imagine, for instance, transplanting a complete head[19] or "more simply", only the brain on a compatible

---

[18]Partial graft: Duvauchelle, Dubernard in 2006; total graft: a Spanish team in 2010.

[19]This has already been temptatively performed on monkeys in the seventies but now Dr. Sergio Canavero, an Italian neuro-surgeon, annonnces it could be performed on humans within two years.

body? Here comes the conceptual problem of deciding if we should either graft a brain on a body or the reverse! Who is me?

Of course, a machine would prevent such moods if a brain could be kept alive *in vitro*, connected to an obedient machine! We then outperform Frankenstein! We have now reached the stage of a drastically transformed man.

Man/machine hybridization

Would machines be the ultimate stage of our evolution?

The hybridization with machines is still "in its infancy" even if the prostheses are largely used. However, the basis of the transformation is already there for actions which could lie outside the experimental framework. Hybridization, although it is still partial, is to undergo a dramatic rise, at least as fast as genomics.

The artificial autonomous heart we were dreaming about until the first heart transplantation in 1969 just became a reality with the "Carmat heart" in December 2013. It was only an incomplete success as the terminal patient survived two months and a half, but he was the first to accomplish such an exploit, as Blaiberg did, in his time, with the historical first heart transplantation in South Africa.

Then the odd question arises: "How does one die with an artificial heart which is guaranteed to be trouble proof and cannot generate a blood clot?" Why can't the other organs not accept this foreign object however compatible? It is for certain that *post mortem* analysis will bring valuable information[20] which the following patient will benefit from. That is the way Science evolves.

Then the machines begin to take over the human and address his failures. From the "exoskeleton"[21] to the brain implant or the brain assisted command, the machine goes for it. When the cohabitation, the connection inert/living, will be correctly mastered this hybridization would no longer encounter any major technical obstacle.

The difficult problem of the nervous connection is not limited to the unique machine; it also exists in the domain of the living. The "classical" heart transplantation is not accompanied by the restoration of the link with the natural regulatory nervous system (ortho and para-sympathetic); the heart frequency then is almost fixed and only stimulated by the sinus node which is kept in place after the graft.

One can raise the following question: When so many opportunities appear for nearly accessible changes, how would man, as an individual, react? Assuredly given the huge diversity of humans with respect to their sensitivity, their culture, their intellectual level, and their beliefs each one

---

[20] According to last information, there are indications that the heart was correctly accepted but the cardiac arrest would be due to a malfunctioning of the electronics; this is very unfortunate for the patient but reassuring for the future as a rapid solution will be provided.

[21] The exoskeleton experiences many developments civilian (Pr Sankai in Japan) or especially military (D Ichbiah, "Génèse d'un peuple artificial", Minerva Ed., 2005).

will react according to his own way, an Indian differently from a German, a worker differently from a doctor.

If it is abruptly proposed to someone to have a microprocessor implanted in his frontal lobe in order to improve his IQ, the answer would certainly be very reticent (especially for people whose IQ is low and find themselves very well that way!). The post-human is certainly not to be accepted by everybody and would remain utopian on a large scale (at least at the beginning). Of course, there would even be people ready to accept anything; everything can be found in the human species! But in any way what is not accepted here, will certainly be elsewhere, without any difficulty!

On the contrary, there is no limit to the creative innovation, no reaction, no principle, no mood when pure machines are concerned; the machine will do what we want it to do. The cohabitation of the machine with humans is only a matter of the external aspect, of form to be respected, but not any *a priori* obstacle is to be found.

The humanoid machine, however, has no reason to be a "copycat" of a human to be more performing. Remember that the bird and the aircraft only have wings in common; no one would actually think of putting a GPS in the head of a bird or put a turbo reactor between its legs![22] Nevertheless, it still migrates over thousands of kilometers as its ancestors did.

---

[22] We just learned that flying insects have been equipped with a microchip to get the control of them and make micro-spies or micro-soldiers.

There still remains the intermediate domain of the "Cyborg" which is a combination of man/machine. We already are accustomed to bear more or less permanent prosthesis: glasses, pace makers, hearing aids, dental implants, artificial hip etc... Thus it would be no surprise. The cyborg puzzle is already there. Artificial blood will soon be available with oxygen enhanced red cells! With the help of GNR the human body becomes reconfigurable on a step by step, discretionary basis.

We have already seen a man deprived of the use of his legs, who runs faster than a good sprinter with the help of his flexible prosthesis! We have already seen a paraplegic woman successfully completing the London marathon, a 452 km trip, by running 3 kms a day with the help of her exoskeleton "ReWalk" built by an Israelian company.

The accessibility to machines by a simple thought (in both the directions if thinkable) will be an important element to bring conviviality between man and machines; this application is in daily progress for some time now. Many things can happen quickly in this domain without calling into question the very nature of man; military are very busy with it and they have done wonders!

The possibility of copying a brain in a machine is a project which is making incredible advances as the brain mechanisms are more precisely decrypted and transcribed in a software language. However, what is not certain is that the final goal might not be as far as to copy a brain but, more likely to be able to simulate its functioning with, may be, a secondary objective to recreate a better or at least more efficient, thus avoiding human weaknesses.

Could it be feasible to outperform the brain? The answer, at the moment, is ambiguous: one would say that it depends on what we want this brain to do; in certain well-defined domains, the machine (the expert system) is already clearly superior, but the brain is also clearly more "plastic" and adaptable. Why be interested in imitating a man if we can directly reach an even better system? Knowing the brain will lead to understand its performances but also its weaknesses as it certainly has. The machine, then, will take care to avoid these deficiencies and the results will certainly mean progress.

To make man cohabit with an "intelligent" machine in the same envelope will therefore not be so easy because each one obeys a different if not antagonist philosophy: man is fundamentally irrational, more sensitive to disorder, impulsive, emotional, fanciful, but not to say unreliable; being rational is in no way a natural trend, he gets rapidly tired but he owns a specific quality which is very difficult to simulate: the power of will.

The machine, for its part, will perform based on a foolproof and tireless logic, a guaranteed rationality, a total reliability, an achieved insensitivity, a defect free memory and a perfect indifference to what makes the human weak: pleasure, anger, jealousy, ... hate! To achieve a viable cohabitation, concession will have to be made both sides, but the machine is rigid and man is shrewd! Who will be the slave of the other? Inevitably this fool's deal will turn into an advantage for one or the other.

Unquestionably, joining man and machine together will follow stages of performance improvements or handicap compensations. Expert system, as they are called, are often superior in terms of specific performances; they count red cells in the blood, pilot military drones, contribute to surgical operations etc... They observe, learn, and memorize the movements to get them optimized; the day they decide they have attained the stature of wanting to go on a war or perform an operation on their own; they certainly will do, especially if they were taught to develop a critical sense and keep into account possible human errors.

In a positive vision we could imagine, a day not so far, that a computer will individually manage our health, it alone would have the means to predict our ills, repair our accidents with a medicine specific to each one of us.

In a negative vision, we could be afraid of the merciless domination of pure logic, but the role of the computer is not limited to that.

In the domain of education, it becomes inevitable that the computer will take an increasing importance in pedagogy (learning is now a word preferred by the software specialists). Instruction, all things considered, is a soft and accepted form of "brain washing" (or "cramming"?); one could imagine, with some fear, what could be possible the day when interactive brain implants would make it possible to transfer "knowledge" in the brain as a simple copy/paste operation.

There are some domains, however, where the help of the machine might be inevitable: the domains corresponding to extreme conditions: especially military and space. Here, every audacity is welcomed. There will be no limit to the experimentation; it will be the "laboratory of the extreme".

Would we put "our fingers in our brain"?

We assuredly know that we all have our own brain, each one different from the other, but, for our good, it is better that our brain be an average of the standard brains. There, however, are brains which are exceptional, "out of the norm", not to say abnormal, those of exceptionally gifted (sometime autistic) whose performances in specific domains as music, language, mathematics are well above average, sometimes, in an inconceivable level.

This (genetic) exceptional performance is often combined with insufficiencies in non-concerned domains or behavioral troubles that reflect deficiencies. This clearly shows that, in each specific domain, the brain could be "boosted" if we knew how to arrange a convenient "wiring" of the brain circuits! The brain does have the biological power to do better.

But nature seems to have a need to compensate such an exceptional development by a compensating limitation elsewhere as if it operates in a constant volume! For a long time, people thought that intelligence was related to the size and shape of the skull, which is not completely true but there is a certain relationship. The "storage volume" or the information is limited and the underlying mechanism of oblivion is to compensate.

We have still to progress a lot in the knowledge of the brain; this science is only 50 years old but the advances are impressive. We now have the required instruments to gather information and allow detailed investigations.

A better knowledge will give us a better understanding why some people are a genius of mathematics or have a gift for languages. It will be necessary to control the operation of the neuron but also discover the global structure and its coordination.

Then it could also be possible to "put the fingers in the brain" in order to compensate the identified abnormalities and then add some biological or silicon Gigabytes; theses prostheses could be internal or external. The approaches, the tracks are diversified but converge on the target: stem cells to regenerate the neurons, elimination of the cellular waste, stimulation through biochemistry or electronic implants.

One already knows how to "externalize" thought to provide with an external control, one has just begun to know how to modify details of functioning (implantation of false memories, images…). This open-door into our "self" is a matter of serious concern, made possible, because of perverse use.

However we have not started from scratch, the attempts to intervene with the brain are ancient and already commonly used in pathologic psychiatrist cases. Active molecules are available from simple tranquilizers to more severe psychotropics not to mention the electrical empirical treatments such as the electric shock in rare and well-identified situations.

Could the rationality of the machine be imposed?

It can be observed that the machines got into the habit of leading their own life, without taking care of humans, and not obeying to necessities or commitments, the reasons for which, we are still pondering. The home PC never asks for our opinion to decide on its own updating; it just informs us that we do not have to bother it during the operation. It is also playing many tricks by itself without letting us understand why, but, in any case, when you think about it, it is always right! Such an insolent attitude of independence will escalate in the future. It is already so difficult to explain to a PC that we definitively do not want to get Google Chrome!!!

The machines soon will know everything about us: Internet social networks, enquiries and statistics of any kind are there to give them every useful information and then they will be intelligent enough to efficiently exploit the data (and humans will also contribute).

One can hardly imagine a machine demonstrating whimsical, irresponsible or perverse feelings similarly to the human. Will the humans accept (more or less easily) to be wrong; the machine never emphasizes such a possibility!

The machine, as any human creation, is to continuously progress in its components, its organization, and its abilities.

The new machines are more often conceived with the help of other machines, increasingly performing and intelligent. To know precisely what an IC is made of and organized, is against the odds and this is to worsen. The rationality in the machine's answer is foolproof. We will need more and more intuition to get accustomed with. The more the machine will interfere in our activities and our lives, the more we have to legitimately fear its dominance because it is always expected to be right. The machine is going to lead "its life".

Chapter 9

Towards a Zero Defect Man?

All these new things will steadily arrive and will upset our lives, by bringing out controversies, diatribes, conflicts of opinion; but the fact is they still exist and we cannot ignore them. So enormous would be the expected changes, we shall have to take them into account.

Bio-conservators versus trans-humanists

We have seen, all along these lines, the arguments and the promises for a surprising future that trans-humanists have made us believe in; promises which are obviously associated to worrying counterparts and patent dangers. This is in no way surprising, life is a continuous question. From our very origins, we are living amongst permanent dangers. Progress has brought us comfort and security along with further sources of anxiety. There are no longer wild beasts roaring at the entrance of the cavern which concerns us, but a nuclear power plant which "could" run poorly. The "conditional tense" here is the most disturbing of all concerns!

The same applies with the announced changes in our extended longevity, our protection against the major scourges, and our knowledge of ourselves. Some see, there, the access to a Paradise others, more suspicious or fearful see the announcement of an Apocalypse (a hypothesis which is not to be dismissed).

In our French microcosm (and the same elsewhere), these heretics belong to all levels of fears: from the soft environmentalists who cannot conceive happiness without a herd of goats on the plateau of Larzac, to the militants who mow the GMO modified corn, vigorously opposed to any evolution and grouped in the order of battle behind the pipe of José Bové. In these revolutionary unpleasant tones we found again the commitment of the silk workers (so-called "canuts") in Lyon two centuries ago. The mechanical looms that Jacquard had just invented! History showed that they were scrambling for nothing; the technology caravan went down imperturbably.

But time is no longer to local and backward looking conflicts, and even if they reach the public stage at a high political level, they lead to laws as fully binding as quickly abandoned. So was it for stem cells in France but also, to a lesser extent, in the US. Realties shall prevail; It is now time for globalization; what cannot be done here will be ineluctably done elsewhere.

As for the interdictions which could be formulated against the implementation of a particular technique or a hazardous product, let us see what happened with devastating drugs like ecstasy.

* From the bio-conservator (or neo-conservator) side:

There nevertheless exists, at a world scale, strong opponents to advanced researches who are narrowly qualified[1] as "bio-conservators" who are petitioning for the establishment of international decisions of regulation or banning for some domains of research they consider harmful or contrary to a basic moral. These extreme positions may result in violent pamphlets.

Among the leaders of this revolt is the spokesperson Francis Fukuyama,[2] an American economist-political scientist from the University John Hopkins (Washington, DC). He was a member of the Bioethics Counsel of the GW Bush administration and he is a follower of Hegel. He uses to preach tirelessly the refusal of future worlds and the protection of our fundamental values. However, he has confidence

---

[1]Neo-conservatism is a political and philosophical complex movement which was born in the US. Its scope is much wider than the context we are dealing with here.

[2]Francis Fukuyama:

— The End of History and the Last Man, Free press, 1992.

— The Great Disruption, Free press, 1999.

— Our Posthuman Future, Farrar, Straus, Giroux, 2002.

— America at the Crossroads, Yale Univ. Press, 2006.

in politics to sort out what could be accepted in biotechnologies and what should be refused, which rather looks naïve.

However, the prospect of notably extending life does not find him indifferent! His only reservation lies in his fear of political catastrophes generated by the blockage of a gerontologist hierarchy.

The grievances developed by these bio-conservators, who play devil's advocate, primarily target the biologists: they logically fear that the acquired knowledge, the neuropharmacology associated to the life extension, would lead to a radical relaxation of our basic notions: choice of moral rules, control of the human behavior, loss of our personality and even our identity to such an extent that world politics and its equilibrium would be threatened.

They consider that genetic engineering, even if it could take time, is a more important event because, apart from its paradisiac promises it puts into question every conceptions we have of our existences.

Epigenetics, assuredly, will make possible to decipher the complex role of the gene sequences of coding (alleles). Even if the hereditary transmission will not be generalized, even if the extension of a genetic modification to the whole humanity is hardly conceivable, even in the long term; it is not mandatory, in a specific scientific context, to have been reaching explanations, before starting experimentation. The point that unexpected or long-term negative effects could appear will certainly not stop the tests. There are fears that we are approaching a domain which is especially unpredictable and dangerous for the human survival.

However, genetic optimization was put into practice for a long time, more or less unconsciously with the choice of marriages favoring the rapprochement of social classes, religious, racial families or equivalent intellectual levels but also discarding kinship or cousinhood.

The bio-conservators point out that, letting parents take the responsibility to "modelize" their progeny can, by itself, be dangerous, their preferences not being necessarily in favor of their kids[3] and the fashion of the day could be disastrous, keeping in mind that to go backward will be impossible. The genomic modification, in principle, is definitive and could sometime be hereditary.

Eugenism no longer remains a state affair but becomes a liberty given to anybody. Religions have also joined the opposition, saying that Man now is to determine his own fate, and not God anymore. Such a lack of respect to the human person, that is a God creature, is considered inadmissible. This attitude is especially promoted by Protestant sects in the US which are still glorifying the Genesis.

The Popes also were in the opposition for the same argument, but Jean-Paul II opened a door by accepting (paying lip service) the validity of Darwin's theory. But he also remarked that the human soul enjoys a privileged status which makes it different from animals. This purely formal reasoning is

[3]Such drifts are to be observed for a long time in the choice of the kid's first names, not even in the best of taste neither in favor of the one which will have to carry it all his life along.

quite skillful and prevents any criticism as the soul, up to now, escaped every attempts of scientific analysis.

The political left in general, especially in Europe, is also hostile to biotechnologies for fear of a possible return of state or class Eugenism. It would not admit that a genetic heredity (intelligence, talents) could be an acquired or transmissible social advantage. It therefore claimed for a state intervention that nobody be left by the wayside in such an irrepressible evolution.

Bio-conservators[4] also fear that the genetic advantages could be kept for the more privileged, thus creating a class of "Gen-Richs" who, because of their acquired superiority, could still stomp out the "Gen-Poors" and could even take power.

However, if the operation becomes cheap because of an extensive application, this could, as well, lead to an egalitarian, uniformized society of "super-somewhat" people. Would it be desirable to build such a hell on Earth? But if the genetic "arms race" would begin, it would be very difficult to escape and maintain the ancient order.

* From the trans-humanist side:

A "utilitarian" trend dominates among the bio-progressists and *a fortiori* among the trans-humanists. James DE Watson[5] defends this point of view arguing that the truth is in the satisfaction of the needs and interests of the society, even if it

---

[4]Lee Silver, Remaking Eden: cloning and beyond a brave New World, Avon, NY, 1998.

[5]James DE Watson, The harm of premature death: immortality, the trans-humanist challenge, document 2009.

cannot be limited to a "cost/benefits" balance. The notion of the "rights" and responsibility remains to be respected beyond that of utility even if there is often a confusion (in the US especially) between rights and interests. The underlying influence of Google (trans-humanism militant) is proving to be more and more determining.

But even the Americans are concerned about the rising hegemonic power of Google. Fortune[6] magazine entrusts its traditional "billet" to the famous columnist Stanley Bing. He evokes, with his usual crazy humor, the near emergence of a successor to the *Homo erectus* that could be dubbed *Homo Googlus*!!!

The bio-progressives argue that modern engineering does not tend to eliminate the weak but most likely help in regeneration to make them strong, the embryonic selection and the genetic technology being in charge of the operation.

Whatever it could be, everybody still remains convinced that the introduction of regulation for the researches, the establishment of "red lines", the implementation of legal barriers may never be of any use due to the speed of the evolution, the number of interlocutors to be convinced, the complexity of the problems where it is very difficult to distinguish what could be potentially dangerous from what could be beneficial.

The field is kept open for the philosophy of constructivism which preaches that everything can be built, including man, and which may lead to a real metaphysical war with a dangerously totalitarian nightmare.

---

[6]Holy floating enigma, Bingman! Stanley Bing, *Fortune*, March 17, 2014.

The biological man and the computer

The biological man we have known until now is to face challenges for which he is certainly not prepared, at the very least, a vast majority of men. Incredible advantages are promised, tremendous threats will appear, and irreversible transgression must be accepted.

A very active domain in software science is devoted to the decision-support tools that already applies to many sectors of the human activity, to the point that it has become essential (economy, industry, financing, transports and military of course). But this technology is to reach lesser traditional sectors such as administration, politics and justice (also may be to reduce the courts congestion[7]?)

The theoretical tools of this branch of applied mathematics involve highly advanced notions as probability, fuzzy sets, graph's theory, operational research etc...They give rise to decision software which draws on information from sources and data bases such as Internet. This is at the moment only a pertinent help but ... closer and closer to the decision.

It is already possible to forecast events from predictive and probabilistic analysis through masses of data: epidemics, suicides...and even the crime,[8] before it is committed!

---

[7]More rationality and less human fantasy could be wishable in the domain of justice.

[8]Police services in Chicago, Memphis, Los Angeles still experiment such software.

Our societies, more and more managed with computers, will experience more and more limited domain of liberty and will be obliged to abandon human methods of management, deemed too unreliable, in favor of more rational and responsible attitudes.

Can we trust in the Man to collectively assume his cohabitation with the machines? We might be doubtful as in many circumstances as he behaves quite incoherently. One can say, as a form of reassurance, that "man is a rational animal" but is it not a presumptuous assertion here? We would perhaps have to wait for the "improved man" or the "zero defect man" to reach wisdom; but today, one simply has to turn on a TV set to understand that the way is longer than expected!

Electronics and biology: same fight

Software assisted neurosciences are in a very fast progress, although operating on "trial and error". The structures of the nervous system (the "hard") now begin to be understood in detail, its functioning too, although the global "orchestration" rules still need to be discovered. We still have no idea of the fundamental point of the learning method: the baby, at the beginning, enjoys the smallest software (one would say a "Bios",[9] in the geek language) which cannot even coordinate the movement but uniquely ensures metabolic vital reflexes. The rest, even breathing, comes from learning on contact with sensations and aggressions from the external world. This

[9] Basic Input Output System.

programming which takes place right after birth is a very long process of accumulation and repetitions; it would take years to make a man; will he ever be achieved?

Nevertheless, auto-educative robots are being tested to understand the human mechanism (in a reverse way from the robotics which proceed from man to make a machine). We still are in the infancy stages with somewhat approximate drafts. One tries to implement rigorous methods in an approximate universe; so much it gets stuck even after using fuzzy or chaotic logic! However, the day is not far off when the rigorous and determinist logic of the Silicium will be able to appreciate the real world and how much efficient the robot will be!

In doing such introspection, we are in some way in the situation of a "caveman" who just discovered a computer and tries to understand how and why it works. We do have our work cut out for us!

In the opposite way if the game is to make a "computer brain", would it not be better to keep it forming by itself, as a baby does? One could imagine an empty computer only equipped with a dedicated "learning Bios". This computer put in relation with the external world through artificial "senses" (which it would have to educate) could acquire its own experience of what is good or bad (that which works and that which do not; the criteria being to be appreciated).

As the baby does, it could be able to get a "conscience" of the world and even share it, compare it with other computers subjected to different constraints, until it becomes a 'brain equivalent' with additional memory, the rigor, the "dispassion" of a machine. I never find anywhere a reference to this naive idea, but which, I think, is as reasonable as that of elaborating from scratch a complete human copy.

All the basis of this idea can be found in Ray Kurzweil[10] who remarked that the brain of an adult man contains a million times more "bytes" than the useful information stored in the DNA. He gave the explanation that it comes from the random auto-organization of external information by a minimum program of the genome; that is to say "life learning".

To do the same with a machine is not inconceivable, if, however it would be possible to associate a "body equivalent" (worthy of the name) which returns the results of the trials.

Towards a "zero defect" baby

This is an essential domain where science convergence and acceleration do triumph. Software (sequencing), genomic (DNA analysis), cellular biology (spermatozoid and ovule), gynecologic medicine (fetus) and obstetrics (delivery) converge in a unique process leading to the creation of a human who, in the future, might well be "produced" in a purely artificial manner. One can hardly imagine the devastating impact, the revolution, that could (are to) be generated in our societies.

[10] Ray Kurzweil, The singularity is near, Penguin Books, 2005.

Google has created a company ("23 and Me") and patented a method allowing select genetic characteristic elements for a baby to be born after an LCD process (size, sex, color of the eyes, immunity to AMD, etc.). They concede that it is not yet 100% reliable but also insure that, unquestionably, chances are improved.

We know that the sequencing of the DNA was on the verge of being a standard operation (with all of the supposed deviation) and at a low cost (now $1000 for a complete sequencing). We know that this DNA could be modified on a discretionary basis and injected in a cell to improve its characteristics. We know that they could give rise to spermatozoids (and soon ovules as well).

We are, then, about to be able, in a laboratory, to "manufacture", purely artificial embryos, with selected, precise specificities and protections. The old way to procreate which required two persons of different sex is, then, no longer required (what a pity!). Everything happens in a test tube and under a microscope. But this still requires proceeding up to childbirth, without a woman's help.

Up to now, to make a baby, two solutions were proposed: either let nature operate and the mother is pregnant for nine months, or implant a newly formed (five days incubation) embryo in the uterus of another qualified woman (the "gestational carrier"). The last solution is called IVF (*In Vitro* Fertilization) and has been largely accepted in many countries after many polemics and extensive discussions.

The following was traditional to lead to the delivery of a new human forty weeks later. This process works and will work, be the embryo a natural one or a "cobbled" or "artificial" one. In this last case, of course, the notion of paternity or motherhood vanishes, and the parents become adoptive. The only trace is the gestational mother who, obviously, is not genetically connected with the baby.

But currently, the technology half-opens a new door: the artificial uterus! This is not yet for real but ineluctably will happen someday.

The issue here is to take over the initial embryo (natural, cobbled or artificial as well) and bring it (him/her?) to the state of a viable human (exogenesis) with the help of machines and synthetic products only.

The idea is in no way a new one: Aldous Huxley[11] already mentioned it in 1931. From that time up to now, and especially recently, experiences were attempted with systems which are in some ways "incubators": the embryo is kept in a kind of minuscule "container" (an adapted shape of 2 × 0.5 mm) where it is immersed in synthetic amniotic liquid provided with endometrial cells in order to mislead the embryo.

Such experiments have been performed in Japan by Yashimori Kuwabara (Tokyo 1996) with goat's embryos (yes?!) and in Cornell University (New York 2002) on human embryos. In the last case, the tissue of the endometrium was grown in a biodegradable cavity, the shape of a human uterus

---

[11] Aldous Huxley, *Brave New World*, 1931.

in such a way that the embryo was not disoriented and has been able to adhere to the uterine wall as it would have done naturally. The growth was done normally for six days, then the experiment was terminated according to law, but, in any case the improvised container would have not been able to extend indefinitely.

At the other end of the chain we found the obstetricians who are doing miracles to save the premature. Current record of the minimum viability in a mechanical incubator is 22 weeks[12] instead of the 40 weeks usually required for a normal baby. This makes it a real performance! Obstetricians would like to be given a "super incubator"; this in order to take care of still younger premature. This machine would have to be connected to the umbilical cord of the fetus through two catheters bringing oxygen and nutrients and dispose waste.

Such an evolutive machine is currently not available to bridge the 21 weeks gap still pending between the primary developed embryo and the viable premature. Henri Atlan,[13] the French specialist, says: "it is quite complex but there is no fundamental biological objection."

However, we have to take care of the whole gestation, that is to say implement a "self-adjusting container" to follow the all development of the embryo; something that the mother's belly traditionally does wonderfully. This is, therefore, a purely technological problem, which certainly will be solved

[12] Survival rate 5%.

[13] Henri Atlan, UA, l'utérus artificiel, Le Seuil, 2005.

one day. This is especially true given that this is not about GNR nor supercomputer to come, but only biology and "classical" instruments.

Another difficulty, may be more important, will be to keep the embryo preserved from infection and viruses as the mother's immunity system will not be there to watch out and protect the fetus. The obstacles for the successful outcome of the project are purely technical, of course, but also deeply conceptual. As easily as it will be accepted when the obstetricians lower the threshold of viability, as difficult it will be to the embryologist to explain to the lawyer to move higher the legal tolerable levels of such experiments. Solution will most likely come from Asia where such reluctances will not hold. We would, then, expect a genuine revolution to come in our way of birth.

It can be taken for certain that, the day, the method and the instruments will be implemented and work satisfactorily (the figure of five years is often mentioned, it is then quite near), an important proportion of women would agree to use it to escape all the inconvenience of a classical pregnancy. Would the health insurances accept paying for nine months in the incubator? That is the question.

Another consequence will be that the woman's role will become limited to give an ovule; men and women would be, in some way, placed at the same level in the procreation process, each one giving his cell, so any affectivity is to disappear.

Of course, this fake uterus would as well welcome artificial embryos. We, then, will have to face this terrific perspective to cohabit with purely artificial "humans" (?) created in a "box" who will mix with the machines of the future. What could happen in this hellish "mixing" of the Man that Nature has taken millions of years to grow?

Will we have to imagine, in this new society, a future Humanity, grown in "factories" of humans as cars are produced today; will there be models we could choose from? Who will take care of these "born orphans"? Who will decide their birth? The sex differentiation being no longer of any use, will we come to non-differentiable and standardized humans? What could be the aim, and for what are we progressing?

Without waiting for these dates, but in a shorter time, many notions will have to be changed. The "random" baby that nature produces today will be "improved" with a series of genetic corrections (often transmissive) to protect him from serious diseases, especially those who are detected as "at risk". Nobody would have the heart to oppose such a precaution which would be more akin to a simple vaccination, at least in the minds of people.

Then the scenario goes further to a second stage where embryos, ready for implantation and issued from a computer modelization, will be proposed which advantageously will replace natural embryos, even the modified. Hence, the genetic filiation will be abandoned for sake of security.

The family context will have to evolve towards a monoparental structure, the life will require. The baby, then, is likely to be like a home pet or a social reference. Love will definitely be removed.

If, at that time, the lifetime is extended, let us say, 50 years "only", that will of course require that the "births" be less numerous and more late. It should be necessary to "regulate the production". May be it will be recommended to wait for the good old days (say hundred years) to turn to raise a child with peace of mind.

Would man still be necessary?

IVF experienced considerable resistance before being accepted. The PID[14] has followed, the genetically "sorted" baby will soon be there, what could happen with the artificial baby?

Of course, the rights of women to decide what happens to her body is perfectly legitimate; it may be thought that the possibility to avoid an uncomfortable pregnancy, lengthy and source of unsightly deformation, will attract many of them.

It will be, as an option, to choose, to keep or modify the heredity or even completely eliminate it in favor of a cleaned, optimized, guaranteed defect free DNA. At the very limit the embryo could be grown from original cells collected on the same individual, man or woman (even barren), or even from synthesized cells. Here are the prospects which are to enter our daily life before the mid-century, not later. All of this is in the only biological domain without any further machine involvement. What do we conclude from this component which will soon enter the constitution of the trans-humans?

---

[14]Pre Implantation Diagnostic allows a selection of embryos to be made, and consequently the choice of the sex.

It is likely that everybody would not feel straight away concerned with such new opportunities, whereas others would readily be.

Cohabitation with the trans-humans will likely be spontaneous, naturally, leading to their fate the ones who do not follow. Same thing already happened in Africa where international industrial companies exploit the mineral wealth, living in comfortable buildings with the blessing of the governments; they went about their business (when they can do) without paying attention to the local impoverished populations which kept on to have babies without taking any care of them and which often regretted the old times of the colonization.[15] This is, here, another way to keep active the natural selection.[16]

An element which is seldom taken into consideration in the discussion is cost, as all we have considered will never be for free. The continued existence of aging people will certainly require costly and careful treatments. In our present society, life is already expensive to maintain due to the fact that man needs to have in his immediate environment, all of the technological paraphernalia nobody could have imagined a century ago (car, washing machine, TV set, telephones, computer, fridge, etc.); he can no longer get rid of them as he could not get rid of his health assurance. What would the situation be like when Cyborg will come with much higher

---

[15]See: http://www.youtube.com/watch?v=dkxySKJcMeY. Kofi Yamgnane's interview.

[16]As long as tribal wars do not contribute with modern weapons.

exigencies? The pure machines alone will not experience the problem of operating cost; they will be self-sufficient as long as they can get enough electricity.

Is Man fundamentally necessary then? Necessary to what, to whom? In which form? These are questions we could readily ask. The danger is that the evolution of our species no longer depends on some individual but rather on everybody, which is largely more risky; individually, Man can be mastered, collectively he becomes uncontrollable and dangerous. The only thing that we must take for granted is that the Earth will continue turning, endlessly, with or without the humans.

* The social impact: towards a senior society.

The immoderate extension of the life, we could forecast to happen in the mid-term, is to pose new problems, some of which already appeared. First of all, mastering aging will have to be synchronized with the cerebral activity management (the decline of Alzheimer) in order to avoid having to care for a population of "immortal dotards"!

As for the hierarchies (in politics, in the business as well as in the labs) we will have to face the presence of 'olds' who will refuse to give up or who have to be intellectually rejuvenated in order for them to cope with new developments. Until now, it was unthinkable to induce a "decline in the hierarchy" because of aging. The only solution was, for a short period before retirement, to give them a "lateral promotion" (in France this is dubbed "the cupboard") towards an honorific but powerless activity. The ideal and definitive solution remains, of course, pure and simple retirement (even if it is a bit anticipated). This chopping is as easy to use as the law is there to set the limits.

Of course, everyone does not hang up to his job or his function; many of them are striving for retirement as soldiers for "demob"!

If the extension of the life becomes rapidly important, the population of seniors (and super seniors) will explode and at the same time the population of young (20–30) will be proportionally weaker. In such a situation it should become urgent to "put our fingers in the brain" not only to compensate the softening of the brain but, really, thoroughly rejuvenate the neurons while keeping the memory and the acquired knowledge.

To complete our conclusions about a life which could indefinitely be extended, would it also be necessary to provide for the ending when the body, after so many repairs, will finally become medically beyond redemption. Would the life-supporting treatments be abandoned or would we have to regulate the euthanasia of these tireless old men (and women too).[17]

All of these conjectures, in the medical domain, ideally suited to the so-called "superior classes" of our developed societies which are intended to be able to appreciate the decisions to be taken (with the corresponding responsibilities) but these hypothesis will certainly be less perceptible for people's minds which are not prepared and for those who do not even ask metaphysical questions.

---

[17] "Die in time", said Zarathustra (F Nietzsche).

And finally one last point: in this evolution towards a permanent medical or technological assistance (both converging) would a class of specialized technocrats on which we would depend emerge? All of that, keeping in mind that the machines could make something of it.

Chapter 10

Is God to Survive Science?

It is understood that touching upon such a subject within few pages is a challenge; the risk is that the subject may hurt people's feelings and acquired beliefs. However, the question remains "is God to survive Science?"; to close this book, some words are still necessary to our reflection, with no partisan approach in mind; this, of course, will not give way to any definitive conclusion.

Science and religion are both involved in their similar yearning to access the light, even if the approaches are quite different: material rationality on the one side against intellectual sophism on the other. What will be the situation given the current scientific advances?

About gods and men

Many of the ancestral fears have now vanished: no more fear of the dark thanks to the electrical lighting; no more fear of the wild beasts, as they have been exterminated (to the point that we now struggle to bring them back in the nature); no more long and tiring walks, thanks to fast transportation; no more hunting for food thanks to the near-by super-market; no more ignorance nor isolation thanks to Internet and social networks.

In return, Man faces himself, the metaphysical anxiety is always there with the appearance of new artificial fears.

What about God in all of that stuff? Would we have to cry out with Nietzsche "God is dead!"?

Until the Man is man, he feels there is an imperative need to set beliefs, seek eschatological explanations, follow commune rules transmitted from traditions, behavioral rituals, and social links otherwise called religion. This happens everywhere and every time because it remains a fundamental requirement of his psychism, an essential need to reassure himself, to get an intimate partner, a hope to clear off the immediate dangers of life, a basic requirement for the group coherency and the tribal belonging, may be the hope of another softer life beyond death.

Religion, in all its forms, is a social requirement, a common and indispensable link of belonging. Some are open to the approach of the others, others are more intolerant, closed on their precepts from the origin until our days, which has led to many bloody conflicts around the world.[1]

---

[1] In the Arab language, the word "religion" does not exist, a more imperative term is preferred, "din", which means the Law. The word "lay" is not translatable in Arab neither in Hebrew.

Every religion is based on a personal choice and is equally respectable as long as it does not call into question the fundamental basis of the society and the individuals. Religion remains, in any case, a necessary pillar at a collective scale; to get rid of it is not so easy even if the sociological changes indirectly issued from the technology ineluctably push towards a certain materialistic and poorly assimilated atheism.

In the past, religions were considered so essential to the society that the children were their favorite target and so emphasis was laid for them to become "well educated". Religion was an integral part of the education in the family. The constraints of the modern life render these obligations more difficult to hold, with the direct consequence of the fading belief over the religion.

The inevitable questioning and transgressions that the future promises us will certainly not make things easier, although Internet and the globalization favor the diffusion of the ideas and the proselytism, not to say indoctrination.

Sorcerers, druids, high priests, shamans, prophets, bonzes or marabous, preachers of all types succeeded one another throughout the ages with more or less success to give their personal version of this notion necessary to the individual equilibrium and the group cohesion with often basic imperative as "thou shall not kill"; all of that was in a framework of a possible rewarding (the Paradise) or a threat of punishment (the Hell); somewhat a religious version of the American version of "big stick and dollars" which still holds as an essential social basis.

In order to win, the religions, were left with no choice but evocate a sense of fear (see Petrone 1st century). This was from the prehistorical ages where the caverns, in prehellenic Crete, for instance, were dedicated to specific cults. Rocks with strange shapes, blistering stalactites in the shaky flames of the torches, immersed the hunter in a worrying atmosphere. Some evocative painting on the walls finished convincing the unbelieving that he has entered another world. Similar dispositions are to be found later in many religious buildings, temples, shrines or cathedrals. Of course deep charges are presently introduced with new churches which look so similar to cafeterias!

At the beginning, divinities were imagined as intimately associated with the nature: hidden in a tree, in a rock, in a river, in the throat of a cave, or deep in the sea, without a means to see them, obviously. All the religions will face the difficulty that they are fundamentally only conceptual.

Greek science and the gods

Historically Greek culture,[2] we have so much inherited of, was directly issued from the poetic invention of Homer or Hesiod and the religion followed the myth, illustrated with a rich mythological catalog born from nothingness. It was in no way a theological construction in the present meaning.

---

[2] Walter Burkert Griechische Religion der archaischen und klassischen Epoche Stuttgart, 1977.
French translation:
La religion grecque à l'époque archaïque et classique — AJ Picard Ed. (2011).

It absolutely does not intend to explain or convince but simply to enchant by the tale and the charm.

Gods, without being visible, but being ubiquitous in the minds and mores, were very close to people, although of a different nature, in a purely imaginary world; they behaved all the same with identical weaknesses and identical passions. The world of gods was in a direct and perfect osmosis with that of the humans. The religion and corresponding sacrificial rites was part of the Greek society; place trust in the gods and participate in the civic duties, else every failure could lead to severe convictions. Socrates, who, however, carefully fulfilled his religious duties, was accused of "corrupting the Athenian youth through his teachings" which encouraged reflecting beyond the sophism, and finally he was to drink the cow parsley.

The Greek religion in no way intended to explain the origin of the world, the cosmos or the immanent nature of man and his destiny. It was a religion for sophists (poets), purely romantic, it could be said.

But Greece was also the birth land of philosophy (philosophia: the "love of wisdom") which pretended to think in a logical way by observing the surrounding world; the conclusion deduced in that manner would superpose or mingle with the religious beliefs. The consequence of this exploratory process ineluctably leads to what we call Science, in an inevitable conflict with the primitive religion. However, the initial intents were the same: establish a truth and stick to it. But the methods of approach were, from scratch, opposite.

With the antique Greece, for the first time, a structured scientific thinking appears which tries to be organized, rational, methodical and supported by a structured mathematics. One no longer does anything but think about a general and universal method. The "laws" that intellectual procedure brings on, lead in the practical domain to what we will call later technology.

For instance, mechanics (levers, pulleys, screws, pumps ...) was previously considered empirical but then gives rise to theories. One no more improvises, one forecasts, it is no longer a matter of beliefs but of realities. Science was born, putting the theories into practice (applications). One leaves the religious dreams to tackle an organized, scientific reasoning[3] of the observation which brings immediate improvements in the daily life (particularly in the art of war which will always remain an essential driving force for science: catapults, ardent mirrors,[4] etc.)

At that time was born the idea of the notion of scientific progress whose positive aspects have immediately been recognized but also the latent threats.[5] The competitive confrontation of this new "truth", with that of the religion was unescapable.

---

[3]Possibly the reasoning is to be purely theoretical and intuitive: e.g., the atomic theory of Democritus.

[4]Genial idea attributed to Archimedes just before dying on a beach of Syracuse.

[5]Sophocle, Antigone (441 AC).

The theoretical atheism then appears with Protagoras or Critias as a possibility worth of a debate. Protagoras obviously was brought to trial, he fled but drowned in a river! But such movement of thought cannot be stopped, once it is launched.

One has to wait until Euripides to see the evolution of things to get a natural adaptation. Gods definitely stay away from the humans to become pure spirits; the idea of the individual and immortal soul emerges clearly. The divine and nature integrate in a new synthesis; it all comes together with the idea that the concepts of the poet only are accessible through philosophy. A twist that calms things down! Platoon[6] is at the very origin of this new theology and he would go as far as to say that sophists are no more than charlatans.

Mathematics, then part of the philosophy, is to play a decisive role by demonstrating that its laws, immaterial and universal, apply even to gods. The philosophy and science, facing similar struggle, will be indivisible to give rise to the new concept of rationality sometimes incompatible with the religion, but here a common ground will be found.

The conflict to come

It has been observed so far that the conflict between religion and science is a long standing issue and there have not been any improvement, on both sides, in the initial motivations so far. The difference between yesterday and today stays. Yesterday, the conflict was only a fight for ideas, for a liberty

---

[6]Platon, Les Lois, e.g., Ed. Les Belles Lettres, 1976.

of thinking, for a free reflection on the physical world surrounding, without any implication on the way of life; while today, precisely, our way of life, not to say the very nature of Man, is to question, in counterpoint, the idea of a creative God.

The preferred vehicle to persuade the basic man to prostrate before a god has been, at all times and in any places, the images; that is what better "talks" to the human mind, especially in the old times where images were rarely to be seen. The images (or sculptures as well) then carried a powerful magic, supernatural, incantatory, suggestive[7] value. Shadows of hands or animal painting in the prehistorical caves, sacred icons, sculptures and holly painting in the less distant religions, evocations of unknown threatening or idyllic worlds, everything is an opportunity to generate fantasies.

However, some rigorist religions like Islam consider that the power of the images is so devastating for the mind that they prefer to be deprived of and therefore prohibit any physical representation of the living world. Christians also did not escape the problem in these times with the quarrel of the "iconoclasts" and the "iconodules".

In the middle ages, Cathedrals were invented with their impressive soaring towards the skies. Multicolored glass panels and marvelous colorful wall paintings (which, since,

---

[7]One can keep in mind the panic triggered in the audience at an early film showing of the Lumière brothers where it was shown a train entering the station of La Ciotat!

disappeared) embodied the central concern of the preacher in his pulpit, in order to give them consistency and print them in the minds. This dazzling spectacle, then, benefited a powerful pedagogical impact on the simple peasants who sometimes came from a long distance to see the marvel. Television did not yet exist to influence the crowds!

However, with the modern ages, photography and cinema have arrived, soon followed by globalist television and even telephone. This makes it impossible to escape the permanent bombing of the images.

Images, thus falling into a trivial banality, lose their magic and incantatory power; they no longer attract the gaze. The TV itself lost its efficiency in psychologically convincing people, at least, who still enjoyed an open mind.

Greeks, always them, were to compete with gods. Daedalus was intended to make marble statues which were so real that they could come to life. Prometheus did the same with clay but he was punished, whereas Asclepius tried to resuscitate the dead and was struck by a lightning thrown by Zeus! The competition with the gods to get an explanation does have a long history but the question remains: is the problem of God be limited to the mystery of the world origin?

In the following period, Science and the observations each time moved back the established precepts. Today the Higg's boson, may be, will open a small window towards another universe, in pursuit of an inconceivable God who in any case could not belong to our materialist world.[8]

---

[8]Jesus Christ already said: "the kingdom of my Father is not of this world!"

Obviously, we are locked in this material world. The theoretical physicist, as well as the mathematician, when they are led to a new abstract immaterial concept is forced to create an evocative reference from the real, to help understand what they mean: waves, cords, tunnel, hole, black matter etc... Science, up to now, never brought a light on a possible "beyond", a possible reality which would have escaped our senses. We do know now that we do not know!

Genesis warned us to keep away from the Tree of Knowledge and the Tree of Life. We do have, torn off their leaves both.

In the occidental world, the Book's religions, unlike Buddhism[9] which has no god(s), have organized their beliefs around a unique God who is at the origin of everything, thus giving an interpretation of every mystery: protective father, world's creation, man's promotion, origin of the supreme truth, immortality of the soul etc... however, entering into precise details of the imposed explanations was, sooner or later, to expose its flanks to criticism.

This spiritual religious framework remains however at the basis of our societies even if it is unacknowledged with its accessories: the people's moral. Of course, this idyllic picture did not fail to create serious drifts. Until today massacre and bloodsheds never stopped on behalf of divine beliefs; it was an ultimate justification.

---

[9]Buddhism consists more in a life philosophy than in a religion strictly speaking; it cohabits without any trouble with other religions as Catholicism (in Japan especially).

However, in their pedagogical effort, these religions were all, more or less, obliged to impose stereotypes, fundamental images, simplistic theories which do not suffer any dispute. This respect of a dogma is essential; otherwise, the entire edifice has a risk to collapse with criticisms, thus losing all credibility. One can hardly imagine for an intellectual straight-jacket to be sustainably imposed in our materialistic world; but could we live without?

One can remember Galileo, before his judges, pretending that the Earth was round and rotates around the Sun (which was generally a recognized truth from the Antiquity,[10] but was forgotten) or Ibn Sina (Avicenna) seeking, with difficulties, to explain to the Caliph that his medicine was not directed against the wills of Allah. Already Thales in the 4th century before JC pretended to give a logical explanation of the natural world!

In the old ages, in our countries, Science often was consistent with the religious culture and everything was told in Latin, the language of the scholars. The migration towards a vernacular language took time to be accepted. In the case of Muslims, the Koran, tirelessly chanted in Arab, was supposed to contain all truths and sciences and hence was never opposed.

Napoleon, during the war in Egypt, questioned an assembly of dignitaries if the Koran did contain any indications of

---

[10] 400 years before Christ, Eratosthenes even succeeded to measure the diameter of the ball with an astonishing precision using a simple stock (*gnomon*) planted in the ground, one day of a summer solstice.

how to cast cannon; the response was yes, without an hesitation; it only requires that Allah gives the inspiration to an attentive Muslim or even an unfaithful, if Allah wants it so. The loop was completed; everything can be justified with rhetoric!

The theory of evolution was also fought with all the necessary strength[11] by the religions and was forbidden, not that long ago, in some regressive countries. There were rearguard actions the physical evidence should inevitably prevail. Somehow the religions succeeded in filling the gaps and continued to protect a necessary and accepted by all morality which was to rule our collective lives.

What could happen now when Science allows investigations which are to simply upset the foundations of the certainties most deeply rooted in our minds; whereas technological advances call for transgressions[12] steadily more fundamental to impose a complete reconfiguration of the human. Would the mind of the average citizen have the strength to adapt?

Today, the research for the Big Bang has overshadowed Genesis. Genomics allows an intervention on the very nature of the man. A new domain of science has even recently been opened which is called "neurotheology" midway between science and religion.[13]

---

[11] If Man is coming from the monkey, then it cannot be so for Jesus who is a man while at the same time the Son of God.

[12] L'ultime transgression, Jean-Pierre Dickès, Ed. Dechire, 2012.

[13] Dieu, ADN et dépression, Laurent Alexandre, Le Monde, November 11, 2013.

Following this theory, the religiosity would accompany or predispose to depression or anxiety; it would play a direct role in the "plasticity" of the brain.[14] Mystical ecstasy can be detected by MRI and a specific "God's gene" would have been discovered[15]! It is however well established that the obscure ambiance (see above) of the ancient churches and the infra-sounds of the church organ- pipes, as the images in the prehistorical caverns stimulate a cerebral activity of meditation and contemplation close to anxiety. A strong disapproval for these researches is to be noted in the Catholic communities but not among Buddhists, as the Dalai Lama readily accepts these innovations of the "cognitics".

The way things are to be appreciated could be quite different for a scientific person who is supposed to know (in principle) what he is speaking about, for the average Boeotian who is to receive a schematized, oriented, truncated information and for those numerous for who this information is of no interest at all. In spite of the efforts for an outreach and pedagogy, a gap still remains between these two last categories of populations which continue to face the same existential questions.

Rationality, until Protagoras, as already discussed, hardly enters in the religious framework; how would the computer be able to handle the Bible or the Torah? Google ignores God!

---

[14]Introduction à la neurothéologie, Camille François, http://www.les-mutants.com/neurotheologie.htm.

[15]L Alexandre, www.lemonde.fr/sciences/article/2013/11/11/dieu-adn-et-depression; aussi: http://rue89.nouvelobs.com/2007/05/24/le-gene-de-dieu-revelation-ou-heresie.

The science and the faith

Other than the way the Creator is implicated in the repercussions of the progresses of the sciences, an important point is to be still commented upon: what is the impact of the Science on the way, in the different approaches, the Man perceives God, that is to say his Faith?

Obviously, this impact depends on the origins of this faith and the social and familial context of education; following history of the men who adopted this religion and following the persuading power of this religion.

Christianity, in its various representations, is in terms of numbers, the more largely spread religion in the world. The "knowledge" always generated fear and mistrust among the Christians. The great scientist Gerbert d'Aurillac was treated as a sorcerer, as the extension of his scientific knowledge was impressive (that did not prevent him to become a Pope).

Some people today see "the hand of Satan" in this crazy technological evolution and rather trust more in the coming of the Antichrist than in a Golden Age of the "improved" man. Transgressions did not stop since the famous Paradise Apple. Consciences and faiths are shattered before this necessity to reconsider everything in the light of the science's progresses, hard to circumscribe. Even though, 84% Americans who spearhead modernity proclaim themselves Christians!

Judaism only implies a limited population of about 14 millions of believers who are scattered all over the world. The implication, the contribution of these believers in the intellectual science is clear for centuries. The rational quest for

God is the subject of lively discussions, although the very essence of God is admittedly unknowable.

Nowadays, with centralization of the confessional state of Israel, this scientific implication also largely spreads onto the technology domain. Jewish are directly involved in the science upheaval, their contribution is essential. This however would not sensibly influence their faith because their religion is not directly impacted; rationality[16] is able to cohabite, within limits, with the Bible for a majority of believers at least.

Islam for its own part fundamentally rejects the scientific reasoning and even the Aristotelian logic which they consider as blasphemies (Allah is the only master of our destinies; it is illusory to uncover his intentions which only belong to Him). This religion directly issued from Arab tribes of rigorist Bedouins largely diffuses as a conqueror towards other peoples less sensitive to this rigor, but never contributed actively to the progress of sciences, contrary to a largely spread conviction with the help of our historians.

The role of the Muslim's world[17] in the history of sciences nevertheless was rather important although unpremeditated. They play the role of promotion and diffusion of the Greek thought along centuries (some said they were "pollinators"!)

---

[16] Les religions meurtrières, Elie Barnavi, Flammarion, 2006.

[17] Le mirage de la civilisation arabo-musulmane, Jean-Pierre Fillard Col. Mémoire d'autrefois, Document Cercle Algérianiste de Montpellier (2006).

through many copies in the Arab language of translations,[18] exclusively made by others (mostly Christians).

Then, Science touched the Muslim's world during its golden age (7th to 12th centuries) and was installed in a long sleep because there were no authentic roots.

Today, the triumphing technology of the modern world cannot be ignored; it is installed as a domineering power all over the humanity through the ways of life it superimposes over our traditions. The numerous problems it generates, in every domain, cannot be squeezed by any ideology, even rigid they impose upon us and we will have to live with them (except an Apocalypse).

It is as such more than, on the theology angle, that Islam will have to adapt[19] and this looks not to be an easy thing if we consider the violent convulsions which stress the Muslim's world today. The original Islam intimately mixes the Islamic law with its pressing precepts and the individual life. This leads to a power of conviction extremely strong: as a testimony of that we could remember the conquest of the Americas by the fiercely Christian Spanish which managed this conquest in a pure Islamic tradition, that is to say with the sword ("bessif") and the God's banner.

In the Islamic religion, the sacred commandments are, then, imperious, definitive and non-negotiable. But the pressure of

[18] Les Chrétiens dans la médecine arabe, Raymond Le Coz, Ed. L'Harmattan, 2006.

[19] Faith and Power. Religion and politics in the Middle East, Bernard Lewis, Oxford University Press, 2010.

the modern life is irresistible, especially the unavoidable social networks, Facebook, Twitter, and others which insidiously dug themselves up to the Saudi society, the ultimate bastion of the Muslim orthodoxy. Would the death of Islam be programmed at the end of multiple and bloody conflicts all over the planet, or would its domination destroy Science? That is the very question today

The mutation towards trans-humans

God is going to step aside and many will lose their reference point (if they still have one) which guided and sustained them till now. Many concepts will have to be reinvented. Will Man (some of them at least) be again an animal guided by his primitive instincts or will he mute in a machine provided with the only rationality?

God sees everything, God knows everything, it is said, but now comes the question: will the Big Brother sit on the right of God, because it too knows everything about everything?

Many attempts, in the history, were made to "socialize" religions with rationality (cult of the Supreme Being, communism etc…), all failed because of a lack of the essential: the enchantment, the spiritual, the inspiration, the fairy tale!

Is God still necessary to provide coherence to men, now that man pretends to control diseases and even death? Will trans-humans still need the idea of God?

Is Man of any importance, does he have a predestined role in the living world? Some people believe it, others doubt, some even think that Man is only a puppet obeying to a superior will, an unknown program. Even the soul, an important component in any religion, is still hidden behind the more accurate investigations of our cerebral activity, without letting us identify it.

The equilibrium of our future would then be very difficult to balance: we are on the verge of dropping our crutches without being sure that we are able to walk!

As I have been promising you, I am aware that I have been asking many questions than I have brought answers; answers that the future will soon afford, one after the other; be patient… and vigilant, we will have to face a troubled future… but, at the same time so promising!

Conclusion

The Crystal Bowl

At the end of this discussion what could we hold concerning our future? Let us take our crystal bowl and have a look inside.

We are to discover that there is a striking resemblance with what happened some billion years before. At that time the ancestral "primitive soup" randomly stirred up molecules which tirelessly assembled or disassembled, following electromagnetic radiation fluxes and electrical chocks, thus creating transient or more stable chemical structures such as amino-acids which can still be found in the deep space.

Then this random chaos generated a "monster" which we call LUCA, the first organized cell, that is to say a microscopic, mythical being; it was provided with a new property: life. Would this birth be due to a pure chance or would it be resulting from a superior will[1] no matter, it, indeed, took place and everything changed.

[1] I was just to write: "Only God knows!"

Millions years accumulated in order for Man to appear among a multitude of species. Was it inevitable? Was it a premeditated will? Is it an achievement? Nobody knows but the thinking human being is there and it is inescapable.

Today, Science develops imperturbably, launching pseudopods in every direction which interleaves to mutually reinforce and to progress farther. Is this new statistical chaos, in his turn, bound to be at the origin of a new "monster" of which we have not yet, by definition, any idea? That is what trans-humanists pretend.

This is not unthinkable if not likely. However, a major difference exists between the ancient and present situation. During earlier days it took million years to get a change, but nowadays some little years are enough to upset everything. We have entered a "temporal compression" that cannot be denied. Everything will take place quickly and we (or our children) will be witnesses, if not to say, even the actors of the changes.

A cessation of the Science progression is hardly conceivable if we put aside an astronomic unexpected catastrophe which could destroy the life on earth. However, in the curse of History many events which were considered impossible have arrived. It could also be possible that a return to fundamentalist and retrograde religious requirements could slow down or stop this evolution thus keeping us back to the times of the "oil lamps" and tribal fights, over some centuries.

We can scarcely imagine such an extreme eventuality but it cannot be rejected outright because it already involves a large number of humans. Malraux said years ago, "the XXIth century will be religious or it will not be!"

Among our forecasters on all sides, some bio-progressists are confident of the future and announce the times for an extended life without age-related diseases, whereas other bio-conservators are concerned and warn against an uncontrolled evolution.

Effectively, the spillover effects of an exploding science might worry; the clumsiness or the disastrous diversions can be feared; the possible errors in the judgment can be dread. But in the counterpart, we can expect that wonderful remedies will be found to our ills; we can have a glimpse on what we are made of, we can reach more comfortable conditions for a new life, we cannot refuse to postpone the fatal issue. Would the machine is to impose its rigor or will we be able to convince it to adopt a few whims?

From the beginning, the laws of the natural selection relied on the diversity of individuals inside the same species; when a category became unadaptable it was surpassed by another one; the diversity was there to bring solutions without for the involved species to disappear. But now the humans decide to do as it pleases them and it could be feared that they ossify their realizations on optimized but standardized models; which would likely considerably reduce our hopes for sustainability.

And also there is the machine; its closer ties with man are evident; to what point and under what form this "hybridization" would take place? We have, yet, no idea of it but, surely, it will take place! Google did not exist some fifteen years ago but today it launches initiatives in every direction and is now head of the innovation. What other Googles are to emerge within the next fifteen years?

In any case and whatever be the opinion of the world deciders, an evident global reality is to be confronted in a very near future. The excessive development of the populations, especially the poorest and the less privileged adapted to the changes. This so much vaunted growth shall inevitably be mastered at one time or another.

But, similar to Icarus who trusted in his feather and wax wings, there is a great danger for us to be burned at the Sun of our Science.

For more information you may visit: http://www.singulariteavenir.fr.

Index

Printed in the United States
By Bookmasters